Is anthropomorphism dead? Most modern students of animal behaviour think that it is and are therefore confident that their own use of anthropomorphic language is purely metaphorical. John Kennedy's point is that explicit anthropomorphism was indeed well-nigh killed for such students by fierce criticism from the radical behaviourists, but that we have to recognize that today there is a new anthropomorphism which is much harder to avoid because it is unintended and largely unconscious. It exists because of a fact only hinted at in the literature: that an anthropomorphic 'fellow-feeling' toward animals, especially higher ones, is built into us by nature and nurture. For that reason, even those who, if they were asked, would firmly reject anthropomorphism nevertheless unwittingly slip into it from time to time. The book provides ample documentary evidence of mistakes traceable to anthropomorphic bias. The final chapter outlines things we can do to minimize the damage done by anthropomorphism to the causal analysis of animal behaviour.

THE NEW ANTHROPOMORPHISM

THE NEW ANTHROPOMORPHISM

JOHN S. KENNEDY
Formerly Professor of Animal Behaviour
Imperial College, University of London

CAMBRIDGE UNIVERSITY PRESS
Cambridge
New York Port Chester Melbourne Sydney

Published by the Press Syndicate of the University of Cambridge
The Pitt Building, Trumpington Street, Cambridge CB2 1RP
40 West 20th Street, New York, NY 10011–4211, USA
10 Stamford Road, Oakleigh, Victoria 3166, Australia

First published 1992

Printed in Great Britain at the University Press, Cambridge

A catalogue record of this book is available from the British Library

Library of Congress cataloguing in publication data
Kennedy, J. S. (John Stodart)
The new anthropomorphism / John S. Kennedy.
p. cm.
Includes bibliographical references.
ISBN 0 521 41064 9. (hardback). – ISBN 0 521 42267 1 (paperback)
1. Animal behaviour. 2. Anthropomorphism. I. Title.
QL751.K44 1992
591.51′01 – dc20 91-24780 CIP

ISBN 0 521 41064 9 hardback
ISBN 0 521 422671 1 paperback

UP

CONTENTS

PREFACE

Anthropomorphism in the study of animal behaviour has been a hobby-horse of mine for more than fifty years. During those years the pendulum has swung both ways between anthropomorphism and behaviourism. I was enjoined long ago to develop my concern with this problem into a book, but I could ill afford the concentration it demanded; and doing experiments was anyway much more fun than writing. Perhaps it is not a bad thing that the book has had to wait so long to be written, brief though it is, since of course my thoughts on the subject have meanwhile been changing, most of all in the last few years. But at the same time the status of anthropomorphism itself has been changing, too. Once a live issue, a butt for behaviourists, it now gets little more than an occasional word of consensual disapproval (and exceptionally a spirited defence). What now gave me pause was doubt that any readers would be found for a book that criticized anthropomorphism. Most people might suppose, nowadays, that a book with that word in its title could only be flogging a dead horse. I have to thank three strategically placed people for convincing me that a book on the particular lines of this one would be worth the effort: Vince Dethier, Jeffrey Gray and Pat Bateson, the last being good enough to read and criticize the whole manuscript. I am equally grateful to Tom Baker, Hugh Dingle, Cathy Kennedy, Peter Miller, Steve Simpson, three anonymous referees and, most especially, my wife, for their backing and comments. I should like to thank, also, Dick Southwood for affording me a post-retirement working haven in the Department of Zoology in Oxford; and Alan Crowden of the Cambridge University Press for his consideration and wise counsel during the publication process.

Oxford, June 1991 — John Kennedy

CHAPTER 1

People have always been very ready to believe that animals are like us in having feelings and purposes and acting upon them. Yet there has never been any direct evidence for this ancient anthropomorphic belief, and some three centuries ago René Descartes broke with tradition by arguing that animals were, in principle, machines. Their behaviour, he thought, could be explained straightforwardly by the material mechanisms inside them. Descartes thus sowed the seed of a materialist conception of animal behaviour. The seed fell on rather stony ground and took 200 years to germinate, but by the 1960s the majority of professional students of animal behaviour had rejected traditional anthropomorphism in favour of Descartes on this point. Keeton spelled out their position at that time:

> "Almost all our words have some sort of human connotation, imply some sort of human motivation and purpose. But such motivation and purpose may have no relevance to the behaviour of other animals, and we must constantly guard against unwarranted attribution of human characteristics to other species. Anthropomorphic or teleological thinking has no place in a scientific study of animal behaviour... English (like all human languages), having been developed around human activities and human interpretations, inevitably reflects these, often with a strong cast of supernaturalism.... You are cautioned, therefore, to recognize the pitfalls inherent in any application of human-oriented language to the activities of other animals..." (Keeton 1967, p. 452)

It was a hard grind reaching this point and the first major break with traditional anthropomorphism inevitably went too far. The new approach, championed by Loeb (1900), his pupil Watson (1930) and Skinner (1938), has since come to be called Radical Behaviourism. Putting it crudely, the radical behaviourists more or less discounted internal causes of behaviour, objective as well as subjective. Admittedly this bald description refers only to the later views of Watson and Skinner and is disputed by Skinner's current supporters (e.g. Branch 1982; Lowe 1983; Amsel 1989); but that was the effective message the radical behaviourists left with most workers. Their school dominated the field for the first half of this century but has been very widely rejected over the last few decades. However, the rejection of radical behaviourism does not mean that the majority of workers have gone back to traditional anthropomorphism, although there has been some regression: "The lessons of Behaviourism have not been lost" (M. S. Dawkins 1980). Today the majority are non-radical behaviourists whom I shall call neobehaviourists (see below, p. 6), and they still take anti-anthropomorphism as axiomatic, something mentioned only in passing. For example: "This is ... merely a covert way of adopting an anthropomorphic posture, a posture that we reject when investigating other aspects of behaviour" (McFarland 1989*a*, p. 132). It has again become a matter of serious discussion that human beings as well as animals may be machines. This was a view that La Mettrie, writing a century later than Descartes but much influenced by him, was brave enough to maintain as Descartes himself had not been. Although nowadays, of course, no one is thinking of machines as simple as the ones that they envisaged, nor for that matter of machines that anyone yet knows how to construct (Gray 1987; McGinn 1987; Penrose 1987; Van Gulick 1988; Barlow 1990). Animals as now envisaged are not the stimulus–response automata which anthropomorphists seem to think are the only alternative that anti-anthropomorphists can offer to animals with minds. "What is needed... is to get rid of the prejudice that machines are essentially simple and deterministic, and to gain an appreciation of the complexity and difficulties in predicting behaviour produced by two or more minds interacting... " (Barlow 1990).

Since it has taken many centuries to achieve the present measure of emancipation from vitalism and anthropomorphism, we may, like Bolles (1975), see this achievement as something to celebrate. It has been a tremendous achievement, something far outweighing the falterings which this book is about. Yet there is no room for complacency. The main point that I want to make is that the scientific study of animal behaviour was inevitably marked from birth by its anthropomorphic parentage and to a significant extent it still is. It has

Empathy without sympathy. (Reproduced with permission. © 1943 James Thurber. © 1971 Helen Thurber and Rosemary A. Thurber. From *Men, women and dogs*, published by Harcourt Brace Jovanovich, Inc.)

had to struggle to free itself from this incubus and the struggle is not over. Anthropomorphism remains much more of a problem than most of today's neobehaviourists believe. But I am calling it *neoanthropomorphism* because the problem has changed somewhat in the last fifty years: anthropomorphism has largely ceased to be explicit and effectively vitalist as it was in the writings of Washburn (1926), Russell (1934, 1946), Bierens de Haan (1937, 1947) and most recently Thorpe (1963, 1965) who was the most cautious: "we can never say that a given piece of behaviour, however elaborate it appears and however much it suggests the presence of consciousness, cannot possibly be the unconscious result of a physiological mechanism.... While, then, we cannot give final proof of consciousness in animals, we can bring evidence to bear which is cumulatively highly impressive and does, I believe, give powerful reasons for concluding that consciousness is a widespread feature of animal life" (Thorpe 1965, p. 474). Accordingly, he took up the explicitly anthropomorphic stance that animal purpose exists at all phylogenetic levels, even the lowest, defining it as "a striving after a future goal retained as some kind of image or idea" (ibid. 1963, p. 3). Likewise Russell: "the objective aim or 'purpose' of the activity controls its detailed course". But the tricky problem now is that neobehaviourists who certainly disapprove in principle of such anthropomorphic thinking sometimes fall victim to it unwittingly. This is not a personal criticism; it was a historical inevitability.

In drawing attention to this danger of the unwitting anthropomorphism that I call neoanthropomorphism I should clear the air straight away by affirming that it is emphatically not my purpose to persuade anyone that anthropomorphic discourse about animal behaviour should be abandoned altogether. This is simply inconceivable for the foreseeable future. Indeed the second main point that I want to make – and it is not original – is that anthropomorphic thinking about animal behaviour is built into us. We could not abandon it even if we wished to. Besides, we do not wish to. It is dinned into us culturally from earliest childhood. It has presumably also been 'pre-programmed' into our hereditary make-up by natural selection, perhaps because it proved to be useful for predicting and controlling the behaviour of animals. It is therefore useful, incidentally, in scientific research on the adaptiveness of their behaviour (see pp. 88–90).

Yet at the same time our penchant for anthropomorphic interpretations of animal behaviour is a drag on the scientific study of the causal mechanisms of it. There is an inescapable ambiguity and inner conflict in the attitude of students of animal behaviour to anthropomorphism. Their nurture and presumably also their nature prescribe it; their science proscribes it. If the study of animal behaviour is to mature as a science, the process of liberation from the delusions of anthropomorphism must go on. The more so, because what we have been witnessing recently is, on the contrary, less awareness of the dangers, with more indulgence towards and even some resurgence of traditional, explicit anthropomorphism; that bodes ill for this branch of science. Those who would have us go all the way back to traditional explicit anthropomorphism are still a minority but they show us the way things could go if we are not careful. They are not all eccentrics who can be ignored. Moreover they are as full of crusading zeal as the radical behaviourists before them: "We have lived for a very long time with the iniquitous view that it is scientifically disreputable to ascribe feelings and cognitive processes to animals ..." (Dunbar 1984*c*). Without going so far, there has been a general drift in that direction (e.g. Cheney & Seyfarth 1990; see p. 91). We are witnessing a new swing of the theoretical pendulum, now back towards anthropomorphism.

This short book addresses particularly, though by no means

only, those students of animal behaviour who are interested in causal mechanisms and are neither anthropomorphists nor radical behaviourists, neither vitalists nor old-fashioned mechanists, but are those whom I am calling neobehaviourists. These are modern behaviourists who differ from their radical forebears in not discounting internal processes in the causation of behaviour (and, often, in not excluding some measure of cognitive activity by their animals if these are 'higher' ones). The grounds for using the term 'neobehaviourist' in this unusual sense are given on pp. 104–5. Most of today's ethologists would come into this category. Ethologists are zoologists by training or adoption, but most of those professional psychologists who qualify themselves as comparative or animal psychologists are also neobehaviourists. Staddon (1989) has recently drawn attention to and deeply deplored the fact that such psychologists are addictively 'anthropocentric', meaning that their aim is to throw light on *human* psychology. The inevitable result (now that radical behaviourism has been discredited) is that they are even more susceptible to witting or unwitting anthropomorphism in their approach to animal behaviour than are zoological neobehaviourists. That is a not unimportant theoretical difference between these two groups of people but apart from this the distinction between them and their theories has become more and more blurred since the time when Hinde's (1966) great textbook on animal behaviour came out with its clarion subtitle *A synthesis of ethology and comparative psychology*. The examples that I have chosen to illustrate my theme are therefore drawn from both groups, although I am much less at home with psychology.

The reader may wish to have at least an outline of the zoological neobehaviourist position from which I personally start out and Chapter 2 provides that. The heart of the book is contained in Chapters 3–6 and consists of nineteen essays on ideas which appear to be erroneous and can be traced to unwitting anthropomorphism. This is not therefore a book with a progressively unfolding theme but rather a collection of essays on important topics in the field, some general and some quite specific. These are extensively interrelated, necessitating frequent cross-references. To start with, Chapter 3 assembles seven concepts that have already been generally recognized as erroneous. Chapters 4–6 deal with twelve further ones on which there is less of a consensus, each chapter dealing with two general and two more specific ideas. Keeton's definition of anthropomorphism (p. 1; cp. Asquith's on p. 9) puts most emphasis on motivation and purpose, but he also gives a stretched definition referring to unspecified "human characteristics" being attributed to other species, an example of which is given, for completeness, in **6.4**. The final chapter considers the constructive steps that we can take to avoid the dangers of anthropomorphism in the study of animal behaviour, while retaining its undoubted advantages.

I would emphasize here that I do not think all the mistakes made by neobehaviourists are traceable to anthropomorphism. Hardly less important as a source of errors is reductionism. Reductionism may of course simply mean looking for the underlying causes of behaviour (see e.g. Horridge 1977; Barlow 1989), an approach which has been spectacularly successful in this century. But the term is being used here in its pejorative sense to mean regarding every whole as no more than the sum of its parts. This is the antithesis of holism (Bonner 1980, pp. 5–8) or "emergentism" as Bunge (1977) called it (see also **6.1**). It assumes that nothing new appears as one moves up from lower to higher integrative levels of a system; and that higher-level events are explicable and predictable entirely in terms of lower-level events. This kind of reductionism has led me into error more than once (e.g. Kennedy 1958). The radical behaviourists were reductionists seeking to reduce all behaviour to simple reflexes and tropisms (the article by Kennedy (1939) is a dreadful example). Moreover reductionism in animal behaviour *complements* its ostensible opposite, anthropomorphism. Because it cannot account for new, emergent properties it opens the way to semi-mystical explanations as in Jan Smuts's Holist philosophy and, to some extent, Gestalt psychology. Physiologists have a powerful tendency to take a reductionist view of whole-animal behaviour because they habitually think in terms of one bodily function at a time (examples in Kennedy 1972). But this topic would require another book and is hardly touched on here.

Another thing I should say at the outset is that I am of course no exception to my claim that everyone remains in danger of falling into anthropomorphism without noticing (p. 32). I can recall having slipped into anthropomorphism four times at least, taking quite a time to realize each slip. On the first occasion I ascribed the extraordinarily persistent locomotory activity of swarming desert locusts to a "locomotory drive" (Kennedy 1951). That was when ethology had just hit the English-speaking world and recent converts like myself were anxious to acknowledge the existence of internal causes of behaviour which we had been taught to discount. Needless to say my tautology advanced our understanding of the causal mechanisms of locust behaviour not one whit. Nor could it, unfortunately, immunize me against further unconscious lapses into anthropomorphism. It is not long, for example, since I was persuaded for a while by Gallup's (1982) striking claim to have demonstrated self-awareness in chimpanzees, an issue dealt with in **5.3**. All I can do is refer the reader to the quotation from Clark Hull at the end of Chapter 2.

CHAPTER 2

2.1 Anthropomorphism and teleology

Anthropomorphism in the context of animal behaviour means "the ascription of human mental experiences to animals" (Asquith 1984, p. 138). We are familiar with three kinds of mental (=subjective, conscious) experience: *feelings* – pleasure, pain, the various emotions and sensations (sense-impressions), *motivations* – the goals and purposes of our actions, and *thought* more or less independent of motor action. We are directly aware of these things only in ourselves, through introspection. This means that simply taking it for granted that animals feel and think too, or explaining a piece of animal behaviour "by merely pointing to the goal, end or purpose" of it (as Tinbergen (1951) complained) are by definition examples of unwarranted anthropomorphism – effectively a throw-back to primitive animism.

There is an important source of confusion here that must be cleared away immediately. It is in the pejorative sense of Asquith's definition above that I shall be using the word anthropomorphism throughout this book. Sometimes, however, the word is used to mean merely pretending for argument's sake that an animal can think or feel as we do. That pretence, the so-called 'intentional stance' which I shall call 'mock anthropomorphism', can be very valuable for the hypotheses it generates about the functions of the animal's behaviour, as described in **5.1** on 'Intentionality'. The vital distinction between these two uses of the term anthropomorphism must always be kept in mind, for it is rather easy to confuse them.

Keeton bracketed teleological thinking with anthropomorphism in the passage quoted in Chapter 1, p. 1. This is because for us, introspectively, the formation of a mental image of the goal, end or outcome of a conscious action precedes the performance of that action and is a prime cause of it (see also pp. 51 and 84). That is what we mean when we describe an action as purposeful, intentional or goal-directed, and it is a human mental experience which we cannot assume that animals have. If we simply assume that they do have it, this is anthropomorphism. Of course, we cannot, alternatively, assert that animals do *not* have any feelings and purposes. There is no direct evidence either way, so Tinbergen (1951, p. 4) took the view that it was idle either to claim or to deny the existence of subjective phenomena in animals, a view some writers still take (e.g. Krebs 1977; Toates 1984*a*; Huntingford 1984).

2.2 Explicit anthropomorphism

However, the question of animal consciousness will not go away. In recent years Griffin (1976, 1978, 1981, 1984), a distinguished and avowedly materialist (ibid. 1984, p. 8) student of behaviour, has assembled a huge array of reports of animal accomplishments and capacities which, he suggests, might be accompanied by conscious thinking. This released a flood of resurgent anthropomorphism under the more respectable label of cognitive ethology (Bekoff & Jamieson 1990*b*; see **5.2**). Griffin advanced most of his anthropomorphic interpretations in a guarded, conjectural fashion, and he conceded that "a behaviorist can argue that a completely unconscious organism could behave in the same adaptable fashion" (Griffin 1984, p. 208). But at the same time many less cautious formulations showed that Griffin himself was convinced that animals do think and feel. For example, "The major significance of the research begun by the Gardners is its confirmation that our closest animal relatives are quite capable of varied thoughts as well as emotions" (ibid., p. 202). He was aware that more conclusive evidence was needed but confident it would be produced soon, especially from the work of the Gardners and others on the acquisition of human language by apes. But it has yet to be produced (see **2.4**). In the last analysis, Griffin's massive case comes down to saying no more than this: if

(Reproduced with permission from Humphrey, N. K. 1986. *The inner eye*. Faber & Faber Ltd, London.)

animals behaving in all those apparently intelligent ways were human, they would probably (though not necessarily) be conscious. Nothing more.

Incidentally, the case is not new. "Griffin's (1976, 1981) plea for acceptance of animal awareness is strongly reminiscent of Romanes' anthropomorphic 'ejective' approach" (Mitchell 1986). "Griffin seems hardly aware how closely this book [Griffin 1976] mirrors the writings of the functional psychologists: James, Angell, Jennings and others, who were active in America around the turn of the last century. They too were interested in the 'evolutionary continuity of mental experience'... Griffin's own interests, his style of argument, his evangelism all have a 70-year-old freshness to them" (Humphrey 1977). Dunbar shared Griffin's conviction that many animals are conscious (see p. 26, below), but he too found Griffin's case weak: "none of his examples is sufficient on its own to provide the kind of rigorous proof that will convince the sceptical rearguard of behaviourism. They will rightly argue that perfectly sound adaptive explanations can be given for behaviour without it being necessary to invoke conscious thought processes.... The sceptics will want to know whether any of their explanations are inadequate if consciousness is left out of the equation, and the answer will by no means always be 'yes' even if animals are conscious in the human sense" (Dunbar 1984*b*). Quite so. Barlow (1987, pp. 365–6) and others have confirmed that last statement by pointing out that a great deal of our unconscious activity (for instance, in performing a skilled task) would look exceedingly clever if it were conscious; but it is not. In fact, learning a practical or social skill means removing the procedures from consciousness (Medawar 1976). Moreover

> "many machines behave in ways which, if they were human, would suggest they had conscious mental

states.... It was just such an analogy which led Rene Descartes to deny the existence of consciousness in animals other than man ... [his] idea of perception sans sensation may seem quite preposterous.... Yet ... recent scientific evidence requires us to take it seriously ... human beings can in fact show the behaviour of perceiving without being consciously aware of what they are doing." (Humphrey 1986, pp. 54–60)

Weiskrantz (1987) refers to some of that recent evidence in a passage quoted on p. 19 below; Marshall (1982) and Gould (1982) quote more of it. If consciousness is not always a necessary part even of human perception and behaviour, then evidence that animals behave adaptably and adaptively is not evidence that they think consciously.

2.3 The behaviourist taboo

Griffin bolstered his case for animal consciousness by arguing that "negative dogmatism generally known as behaviourism" had so far prevented the possibility of animal consciousness from being seriously considered. "Very few scientists," he said, "even realize the extent to which their thinking is constrained by this behavioristic taboo" (Griffin 1984, p. 20). Of course the behaviourists did not literally impose a taboo. But they did manage to convince many scientists that they would be guilty of unscientific thinking, even vitalism, if they described any behaviour in subjective, teleological language. The charge may have been correct in principle, but in practice, although a description might be couched in such unacceptable terms, it often alluded to a real phenomenon that had been overlooked by behaviourist observers wedded to over-simple concepts of behaviour. It is very true that "Fear of the dangers of anthropomorphism ... caused ethologists to neglect many interesting phenomena" as Hinde (1982, pp. 77–8) said; and Kummer (1982) elaborated the point. The neglect was an overswing of the theoretical pendulum away from the rampant vitalism and anthropomorphism of the pre-behaviourist era. As Lorenz (1950, p. 223) recalled, the mechanist and vitalist schools fell into "reciprocal errors". But, he added, "we are justified in regarding the vitalistic errors as primary and the reciprocal errors of the mechanists' reactions justified in themselves and erring only through exaggeration".

Griffin, on the other hand, made the following ingenious suggestion in further support of his case for anthropomorphism: "Why do hard-headed scientists, so anxious to avoid implying that animals think or feel, use terms that in ordinary usage do connote conscious thinking?... Perhaps what the behavioral ecologist observes in nature suggests consciousness so strongly that part of him does wish to suggest that animals think about the likely results of their actions" (Griffin 1984, p. 24). In fact, "hard-headed" scientists (i.e. neobehaviourists) have a less devious reason for using anthropomorphic language to describe animal behaviour than a suppressed belief that animals really are conscious thinkers (see also **7.1**). They choose to use our ordinary everyday language, although it is anthropomorphic, for the simple reason that it is our everyday language and therefore readily understood (Krebs & Davies 1981, p. 3; McFarland 1989*c*). Among themselves neobehaviourists believe it is being used in a purely metaphorical sense and really refers to the functions of the behaviour (Krebs & Davies 1981, p. 256; see also **3.6**). The temptation to believe that animals think about the future is undoubtedly there, but neobehaviourists believe they are aware of it and resist it. R. Dawkins used a most readable anthropomorphic style in his book *The selfish gene* and took the trouble to explain that his uninhibited anthropomorphism was not to be taken literally. "We allow ourselves the licence of talking about genes as if they had conscious aims, always reassuring ourselves that we could translate our sloppy language into respectable terms if we wanted to" (R. Dawkins 1976*b*, p. 95). However, the book owes much of its enormous success to its exceptional readability, its fluent and vivid anthropomorphic style, and this will have done little or nothing to encourage readers to stop and translate, for themselves, the conceptually "sloppy language" into objective terms. Encouraging that habit is still the only way to discourage 'sloppy', i.e. anthropomorphic, thinking about animal behaviour, but it is seldom done (see **7.4**).

2.4 Uniqueness of *Homo sapiens sapiens*

The most widely held scientific reason for assuming that there must be some measure of consciousness in animals is the Darwinian principle that evolution has been a continuous process. Since the human nervous system is only quantitatively different from that of animals, Griffin argues, like many others from Romanes (1882, 1883) onwards, that products of the nervous system such as mental awareness should also differ only in quantity. Since the argument for animal minds depends critically on an analogy with ourselves, there is no discernible limit to how far we can carry it down the phylogentic scale. So, paradoxically, Griffin takes no account of an animal's position on that scale, following Romanes. "Having full regard to the progressive weakening of the analogy from human to brute psychology as we recede through the animal kingdom downwards from man, still, as it is the only analogy available, I shall follow it throughout the animal series" (Romanes quoted by Wasserman 1984). Thorpe and others are more consistent: "the only position compatible with the theory of evolution is that the development of consciousness has proceeded alongside that of the organic structures with which it corresponds (Thorpe 1974:

319–21) and consequently that it is present with varying degrees of elaboration and complexity at least in all higher animals" (Ingold 1986, p. 18). Krebs (1977) produced the short answer to that argument: "This assumes, of course, that quantitative differences are not translated by thresholds into qualitative effects." It seems that such a "translation" is exactly what probably happened in the evolution of *Homo sapiens*, as the anthropologist Isaac believed, below. Admittedly, another anthropologist, Geertz (1975, p. 62), summarily dismissed that "'critical point' theory of the appearance of culture..." as being "conceptualized as one of marginal quantitative change giving rise to a radical qualitative difference, as when water, reduced degree by degree without any loss of fluidity suddenly freezes at 0 °C...". But since then Passingham, a primate psychologist also at home in biology, has put together a body of facts about the development of the size and organization of the brain and the ways in which the brain acquires information postnatally that requires us to take Geertz's derisive description seriously. For example,

> "How could it be that the *mental* distance between a person and a chimpanzee could so outstrip the *morphological* differences between them?... The genetic distance between a man and a chimpanzee is small: in fact, it is smaller than the distance between a mouse and a rat. And the story is similar if comparisons are made of structural proteins.... It is possible for a small genetic change to have very far reaching effects.... Our brain is three times larger than it should be for a primate as heavy as we are.... The device for producing so large a brain is a simple one. We need suppose only an alteration in genes controlling the timing of growth." (Passingham 1989)

"The most dramatic step came in early man" wrote Bonner (1980, pp. 188–9), "...a new and important genetic change occurred that made the progress in this cultural transmission suddenly increase at an astounding rate ... there was a change in a few genes that affected the timing of development ... the brain was simply allowed to grow for a longer period of time than the rest of the body ... a small genetic change produced a larger brain, that in turn was masterly at handling memes in a variety of ways including complex teaching." At what stage the 'translation' happened is quite unknown but the incipient development either of consciousness, or somewhat later of our grammatical language, would seem to be good candidates for the point of 'take-off' by *H. sapiens sapiens*.

Gould (1982), on the other hand, flatly dismissed the idea of any difference of kind between human and sub-human behaviour. His monumental treatise on ethology provides a wealth of documented evidence of 'genetic programming' of the behaviour of all animals. Comparing primates and human beings, he makes a convincing case that we did not "come into the world unprogrammed" (Gould 1982, p. 499) and declares that "the old notion that our species is different from the animal world ... has become more and more absurd" (ibid., p. 483; cp. Dennett 1987, p. 110–12). This is fair comment as far as the reality of 'genetic programming' is concerned but in other respects Gould himself pays tribute to our uniqueness. For instance, "our species seems uniquely driven to make sense of the senseless, the uncertain, or the unknown, while other species face such ambivalent circumstances with placid apathy or generalized fear" (Gould 1982, p. 492). Or again, "We cannot know where, during the course of evolution, our increasing mental capacities spawned the will that now battles with our genes for control of our behaviour" (ibid., p. 541). The three 'our's' in that second statement carry the clear implication that rebellion against the genes is a product specifically of human evolution, unknown in other animals. R. Dawkins (1976*b*, 1989) and Dennett (1987, p. 298) agreed that

> "[human brains] even have the power to rebel against the dictates of the genes, for instance in refusing to have as many children as they are able to. But in this respect man is a very special case." (R. Dawkins 1976*b*, pp. 63–4)
> "We, alone on earth, can rebel against the tyranny of the selfish replicators." (Ibid., p. 215)

Human beings looked like a special case to the archaeological anthropologist Isaac: "we are rightly impressed with the biological success that seems to have followed from the development of the [human] brain through some critical threshold.... The brain is the organ of culture... the intricate body of language, craft skills, social custom, traditions and information which humans learn.... Cultural complexity and flexibility of this kind is unknown in any other organism" (Isaac 1983). The contrary view of Ingold (above), another anthropologist, rests on his gratuitous claim that animal action "is demonstrably governed by conscious purpose" (Ingold 1986, p. 35). The biologist Medawar (e.g. Medawar 1976), on the other hand, made the same point as Isaac many times, and the philosopher Fox (1986) enlarged upon it while Passingham (1989) wrote of man as "a creature quite unlike anything the world had ever known'. It cannot, then, be assumed that the continuity of the evolutionary process means that we differ from other animals in degree only, or that other animals must be conscious to a significant degree.

2.5 Consciousness

In addition to the renewed advocacy of the very ancient notion of animals as conscious thinkers like ourselves, there has recently been renewed attention to the function of our consciousness. The material mechanism of human consciousness is of course quite a different problem and there has been virtually no progress in understanding it (Blakemore & Greenfield 1987). But Humphrey echoed Morgan (1908) in saying: "either we throw away the idea that consciousness evolved by natural selection, or else we have to find a function for it" (Humphrey 1987). He has set forth an ethologists's logical reasons for supposing that its function is to act as the brain's self-monitoring system, "the Inner Eye" as he called it (Humphrey 1986, 1987). Weiskrantz's experiments on brain-damaged human patients with visual defects led him to a like conclusion:

> "What are the implications of these two classes of patients, those with 'amnesic syndrome' and those with 'blindsight'? Both illustrate really quite striking capacities in the absence of the patient's own conscious knowledge. The person can process information if it leads to a straightforward and ambiguous route from stimulus to response, in the absence of 'thought'. What I think has become disconnected is a monitoring system, one that is not part of the serial information-processing chain itself, but which can monitor what is going on.... The 'monitoring system', to paraphrase Lloyd Morgan, is where intelligent processing ends and consciousness begins." (Weiskrantz 1987, p. 319)

Note that Weiskrantz, unlike Humphrey, envisaged not just one monitor but tier upon tier of them. Moreover, these monitors have executive power: "another monitor looking at a set of other monitors... an elaboration of hierarchical levels of organisation, and hence of varying levels of abstraction in thought... action must be capable of redirection as a result of the 'monitoring'; the monitor is not just another of Huxley's epiphenomenal steam-whistles. 'Monitoring' reflects a form of neural organisation with a hierarchical capacity for control" (Weiskrantz 1987). That is an idea shared by Penrose (1987, Fig. 18.3, p. 267), who illustrated it with a cartoon chairman of a large corporation deciding between two sets of highly processed and simplified data assembled for him by minions at the lower levels of his organization.

There has recently been a shift of opinion concerning the selection pressure behind the evolution of intelligence and eventually of consciousness in the higher primates. The main pressure is now widely thought to have come from the growing complexity of their social interactions, rather than from tool-using and other practical activities, as used to be thought (Jolly 1966, 1972; Chance & Jolly 1970; Humphrey 1976, 1983, 1986, 1987; Crook 1980, 1987; Cheney & Seyfarth 1985; Barlow 1987, 1990; Byrne & Whiten 1988; Dunbar 1989; Zihlman

A model of consciousness. (Reproduced with permission from Penrose, R. 1987 In Blakemore, C. & Greenfield, S. (Eds.) *Mindwaves. Thoughts on intelligence, identity and consciousness.* Basil Blackwell, Oxford.)

1989). Given consciousness, human primates were able to communicate their subjective feelings and intentions to their fellows, and moreover, by analogy, to envisage what their fellows were feeling and therefore would do, and to act accordingly.

> "Not only are [human] brains in charge of the day-to-day running of survival-machine affairs, they have also acquired the ability to predict the future and act accordingly." (R. Dawkins 1976*b*, p. 63)
> "Consciousness provides us with an extraordinarily effective tool for understanding – by analogy – the minds of others like ourselves." (Humphrey 1987, pp. 380–1)
> "We, the reason-representers, the self-representers, are a late and specialized product. What the representation of our reasons gives us is foresight...." (Dennett 1987, pp. 317–18)

Humphrey cited the seventeenth century materialist philosopher Thomas Hobbes making just the same point: "Given the similitude of the thought and passions of one man to the thoughts and passions of another, whosoever looketh into himself and considereth what he doth when he does think, opine, reason, hope, fear, &c., and upon what grounds, he shall thereby read and know what are the thoughts and passions of all other men upon the like occasions" (Hobbes 1651, cited by Humphrey 1986, p. 72).

Plainly, such a capacity for predicting the behaviour of fellows and acting accordingly would have thrown wide open the door to cooperation with those fellows to mutual benefit (Passingham 1982), extending indefinitely the scope for reciprocal altruism (Trivers 1971, 1985) enormously facilitated by language. Of course, it will have opened the door at the same time to competition by means of deception and cheating

(Trivers 1985; Byrne & Whiten 1988; McFarland 1989*a*, pp. 126–7, 146–7; *b*). These and other authors devote more attention to this competition and deception than to the cooperation. This may be partly because as Dunbar (1985) believed "the most plausible attempts to show that animals are conscious have come through asking whether they can lie deliberately". McFarland (1989*a*, *b*, p. 146) even went so far as to suggest that "our belief that our behaviour is (sometimes) intentional is the result of evolutionary designed self-deception". Perhaps we find incidents that look like deception to us especially convincing as evidence of animal consciousness simply because we ourselves are intensely conscious of what we are doing when we engage in deliberate deception. But that (an example of unwitting anthropomorphism?) is by the way. Competition in deception, deception-detection and counter-deception will have had the important consequence of driving cultural and perhaps biological evolution through "arms-races" (Dawkins & Krebs 1979), but no society in which communications were predominantly deceptive could last. Crying "wolf" when there is none is notoriously counter-productive. In reality, as Trivers (1985, p. 395) put it, "deception is a parasitism of the pre-existing system for communicating correct information"; or Axelrod (1984), "We all know that people are not angels, and that they tend to look after themselves and their own first. Yet we also know that cooperation does occur and that our civilization is based on it" (cp. Smith 1986; Noble 1989; Whiten 1989).

Our material achievements have obviously depended on massive cooperation achieved by communicating correct information. These achievements are, after all, staggering: "cultural evolution... has converted us into animals that are simultaneously aerial, terrestrial and submarine, processing X-ray eyes and sense organs sensitive enough to feel the heat of a candle at a distance of a mile.... This system of evolution is the characteristic to which we owe our clear-cut biological supremacy over all other organisms because it has conferred almost unlimited capabilities upon us" (Medawar 1976, p. 502). Even more impressive than our material achievements (some of which we now recognize are against the long-term interests of everyone) are our intellectual achievements in the planning and designing of the technological triumphs, to say nothing of our non-technological cultural constructs: artistic, scientific, religious, political, legal, administrative and so on. The plain fact is that

> "human beings have evolved to be the most highly social creatures the world has ever seen. Their social relationships have a depth, a complexity, and a biological importance to them, which no other animals' relationships come near." (Humphrey 1987)

If, therefore, this explosive development, which is thought to have required only some tens of thousands of years but amounted to a qualitative change, was made possible by the development of consciousness *de novo* in our species, then it seems most reasonable to infer that other animals are not conscious. Humphrey went on "No accident, I think, that human beings are as far as we know unique in their ability to use *self-knowledge* to interpret others. If that ability could exist without consciousness, let someone prove it to me. If any other animal possesses it, let someone tell me what evidence he has" (ibid.).

Humphrey was rather more equivocal about our uniqueness in his 1986 book, where he cited Premack's and Gallup's researches as evidence that not only we but possibly apes also "have insight into the way their own minds work" (Humphrey 1986, p. 81). Crook (1987), too, was more ready to concede some consciousness to apes than he had been in his 1980 book. But they both invoke the work on ape language and self-awareness discussed here in **3.4** and **5.3**, respectively, where it is concluded that the results cannot be interpreted to mean that these animals are conscious.

2.6 Unconsciousness

Altogether, then, it seems likely that consciousness, feelings, thoughts, purposes, etc. are unique to our species and unlikely that animals are conscious. If we were entirely logical about it these probabilities would be enough to make us try to avoid anthropomorphic descriptions of animal behaviour. But we are not entirely logical about it, and we have to ask why scientists as well as laymen should be so addicted to anthropomorphic expression. Various reasons have been advanced such as the convenience (e.g. Tinbergen 1951), the ease (e.g. Dunbar 1984*a*; M. S. Dawkins 1986), the appeal (McFarland 1989*c*), the colourfulness (Stuart 1983) or the familiarity (Kennedy 1987*a*; McFarland 1989*c*) of such language, as opposed to the "dry" and "non-committal" (Tinbergen 1951), "long-winded" (R. Dawkins 1976*b*, p. 95), "cumbersome" (Stuart 1983), "sterile and dull" (Burghardt 1985) language to which we have to resort in trying to avoid anthropomorphism. But the fact is that we do not make an impartial, rational choice between these alternatives, coolly weighing their relative merits. That unwelcome truth surfaces in the literature not infrequently, often giving the curious impression that the author is rather embarrassed by it. The following quotations come from an early comparative psychologist, a zoologist, two ethologists, a physiologist and two contemporary psychologists, in that order (my italics in each case).

> "*Whether we will or no we must* be anthropomorphic in the notions we form of what takes place in the mind of an animal." (Washburn 1926, p. 12)

"It can be argued that no matter how excellent and pure our stated intentions may be, the words will unconsciously tend to *make us* interpret animal behaviour in human terms." (Bonner 1980, p. 12)

"...the teleological imperative – that people are *predisposed* to attribute intention (in the narrow sense) and purpose to other people and to animals...." (McFarland 1989*c*)

"when we watch an animal 'searching' for food, or for a mate, or for a lost child, we *can hardly help* imputing to it some of the subjective feelings we ourselves experience when we search." (R. Dawkins 1976*b*, p. 53)

"...there is abundant objective evidence...that lower animals...have lasting phases of different responsiveness from other periods. One *cannot avoid* thinking of moods...." (Bullock 1965, p. 309)

"Humans not only attribute purpose, intent, and various mental states to one another, but there is a primate and *almost irresistible* tendency to generalize such attributions to pets and other animals." (Gallup 1982, p. 243)

"We *cannot avoid* teleological thinking and...it is therefore essential that we should engage in it consciously and critically in order not to be led astray." (Henry 1975, pp. 105–6)

We must indeed be under some kind of pressure to be so ready to believe in animal purposiveness even when it is impossible to meet the normal scientific requirement for proof. Even when authors do their best to avoid anthropomorphism, they find it difficult.

"The study of animal behaviour is unique among the sciences because it begins historically and methodologically with human behaviour, prescinds from human experience, and projects this experience into other

> animals. It is thus more disposed to subjectivity and introspection than the other sciences and constantly labours under the burden of containing these biases within their historical context...." (Dethier 1964)
>
> "Everyday language carries meanings beyond what the users may wish to imply in behavioural research." (Visalberghi & Fragaszy 1990*a*)
>
> "Anthropomorphism arising through ordinary language terminology appears in reports carried out with scrupulous attention to objective methodology... simply as a result of the nature of our language.... Reference to a conscious subject always slips in, whatever the disinfecting precautions, simply because language has been so framed as to carry it." (Asquith 1984)

In other words our ordinary everyday speech is anthropomorphic. It is not neutral and non-committal as Kummer (cited by Harré & Reynolds 1984, p. 107), for example, wanted us to believe. The urge to endow animals with our mental abilities is so strong that it brought great disappointment for the several groups who tried to teach chimpanzees human-language-like skills in the 1970s (see **3.4**). This same anthropomorphic urge seems to have created illusions also among students of primate social behaviour in the field. Dunbar (1985, p. 39) acknowledged that his evidence did not actually prove that animals are conscious but then added dismissively, "although anyone that can doubt it must live in rather a peculiar world". He had prefaced his account of social behaviour in gelada baboons with this declaration: "I shall make frequent use of the language of conscious decision-making in defiance of Lloyd Morgan's proscription of anthropomorphisms [*quoted on p. 153*]. I do so partly because this is much the easiest way, but also because fifteen years of field work have made it abundantly clear to me that strategy evaluation is precisely what the animals are doing" (ibid. 1984*a*, p. 4). He felt sure this view of his was quite typical: "Most field workers – whose research places them in very close contact with their animals [primates] – have no doubt that their animals act consciously" (ibid. 1985; cp. Goodall 1986).

Should we then assume that these experienced workers know better than people who are sceptical about the consciousness but have seen less of the animals? Gould (1982, p. 483) thought so, and so evidently did Dunbar. Yet the considered judgement of de Waal, an exceptionally experienced primatologist quoted on this subject by both Asquith (1984) and McFarland (1989*a*), was different. He described a number of examples of "non-ritualised, intelligent forms of deception among semi-captive chimpanzees" which he said, "suggest, but cannot prove, the existence of *intentional* deception, since many of these examples lend themselves both to complex cognitive and to simpler explanations" (de Waal 1986). He then went on to make a very perceptive remark: "One factor which seems to have some effect on a scientist's attitude about the controversial issue of animal mentality is the amount of experience he or she has had with the behaviour of non-human primates, especially pongids. The fact that absolute 'nonbelievers' [in animal premeditation and intentionality] are rare among people familiar with members of these species means that direct exposure to their actions has the power to convince. Rather than on the gathering of explicit evidence, this process of gaining confidence in cognitive explanations is based mainly on human intuition" (de Waal 1986). Bierens de Haan (1947) defended intuition as a means of explaining animal behaviour but today intuition by itself is regarded as unacceptably anthropomorphic. Intuitions, or "hunches got by introspection" as Bunge (1977, p. 509) called them, "must not be regarded as self-evident but as hypotheses to be subjected to objective tests", as he said. Or note again McFarland (1989*a*, p. 147), referring directly to de Waal's statement above: "A scientist's hunch is acceptable as a start, provided that it leads to a theory that can be rejected in face of the evidence." A cognitive explanation based on no more than intuition cannot be given that Popperian test. It is anthropomorphism pure and simple.

In the event, when discussing whether it is useful for observers to make the assumption that higher animals are conscious, Dunbar effectively confirmed de Waal's point about the intuitive basis of cognitive explanations of primate behaviour. "In practice", he said (Dunbar 1984*a*, p. 232), making this assumption "may do no more than formalize what every good ethologist in fact already does intuitively". Elsewhere, Dunbar even paraphrased Asquith's (1984, p. 143) argument above: "the very language we use derives from human experience and, as a result, it inevitably presupposes consciousness. There simply is no 'neutral' language in which to describe the behaviour of animals that does not prejudge the issue" (Dunbar 1984*c*). Even if this is something of an overstatement, in saying it Dunbar knocked the props from under his own conviction that animals act with conscious intent.

2.7 Compulsive anthropomorphism

Why is it, then, that people persist in taking it for granted that the behaviour of animals is consciously intentional? The answer has been staring us in the face for years, and especially since Tinbergen (1951, p. 4) thought it worth mentioning the familiar fact that "introspection leads us to believe that our own behaviour is controlled, to a certain extent, by 'foreknowledge' of ends or goals". In other words the belief that our behaviour (and by analogy that of some animals too) is goal-directed and intentional is simply *built into us*. Washburn, Bonner, McFarland, R. Dawkins, Bullock, Gallup and Henry all virtually said as much in the passages quoted together above (pp. 24–5). Recently Tinbergen's tentative point has been developed quite explicitly by McFarland:

> "Introspection tells us that much of our own goal-seeking behaviour is intentional, and we tend to assume that the behaviour of other people, of some animals, and even of some machines, is similar". (Ibid. 1989*a*, p. 125).
> "... we are designed to think in teleological terms. This mode of thinking is useful in interpreting the behaviour of our political rivals...." (Ibid. p. 147)
> "We naturally talk and think in teleological terms... we find it very hard to divorce ourselves from the purposive anthropomorphic view, which I propose to call the teleological imperative". (Ibid. 1989*c*)
> Our evolutionary inheritance pre-disposes us to interpret the world in terms of meanings and purposes, as if the rivalries of our political life were relevant to the inanimate world. The result is that we attribute purpose where there is no purpose, and seek meaning where there is no meaning." (Ibid. 1989*c*)

The child asking "What is the purpose of flies?" is not alone in thinking there has to be one.

It must be a mistake to believe that our behaviour is literally intentional in the sense of being controlled by foreknowledge of its end, as Tinbergen put it, since an intended end often fails to come about. Even when the actual end does turn out to be the intended one, this actual end cannot have been controlling or serving in any sense as a cause of the behaviour, for that would require a teleological reversal of cause and effect. On the other hand our behaviour could be controlled by a mental image of some end or other, whether that turned out to be the actual end or not. That after all is what 'intend' means. This process would not be teleological because here the behaviour is not caused by clairvoyant foreknowledge of its actual end-result, but by a mental image or 'representation' of some end-result. The mental representation of an end-result precedes the brain's commanding of the behaviour, so the cause precedes the effect in the orthodox manner.

Now the only datum available to the conscious brain in formulating the intention consists of the learned association between the given behaviour and a certain outcome of it. It follows that the memorized association between them must be reprocessed in such a way as to invert the order of events before it is presented to the 'inner eye' of consciousness, so that the mental image of the outcome comes to precede the behaviour. Such an inversion should be well within the human brain's capacity for signal processing; especially since, as R. Dawkins (1989, p. 141) said, natural selection would favour the development of "our goal-seeking capacity...an immensely useful piece of brain technology". The neurophysiological capacity for such inversion could well have been the key new feature that enabled *Homo sapiens* to 'take off'.

McFarland (1989*a*, *b*), however, has mounted a major attack on the belief that there is such a thing as truly 'goal-directed', intentional, behaviour in animals or even in people. He does not accept the usual view that mental goal-representations play a causal role even in our own behaviour. Here he parted company with Weiskrantz (p. 19) and many others including his fellow participants in a lengthy debate on goal-directed behaviour (Montefiore & Noble 1989). McFarland's argument against goal-directedness is based on the inevitability of frequent 'trade-offs' between the competing behavioural demands on an individual in any real-life situation. This he argued would be incompatible with effective goal-direction, since that would require active exclusion of all other goals from any role in the direction of the behaviour. "Such trade-off is incompatible with the essential design feature of goal-directed behaviour, because active (goal-directed) control systems are designed to eliminate the influence of extraneous variables, whereas the essential feature of trade-off is to allow such influences" (McFarland 1989*a*, p. 123). That may be so in animals, but R. Dawkins on the other hand is convinced that human brains are (uniquely) able to combine "flexibility in setting up new goals, coupled with tenacity and inflexibility in pursuing them.... What natural selection has built into us is the *capacity* to strive, the capacity to seek, the capacity to set up short-term goals in the service of longer-term goals" (R. Dawkins 1989, p. 142).

Moreover, if consciousness makes mental images, and therefore intentionality, possible as I suggested above, then, in the case of people and only of people, the argument that goal-directedness is teleological loses its force. In fact, McFarland (1989*a*, p. 147, quoted on p. 29, above) described our teleological mode of thinking as useful in interpreting the behaviour of our fellows, in agreement with Weiskrantz and Penrose (p. 20). To be useful, our teleological thinking has to influence our behaviour. If it does not influence our behaviour, then we are not using it. And to survive under natural selection it must have influenced our behaviour.

It is not very difficult to imagine what may have promoted the evolution of our in-built empathy with animals. Once we had evolved this empathic awareness of the feelings, thoughts and intentions of our fellow human beings (p. 21), then there could have been positive selection for extending this way of thinking to cover animals, during the hunting and domestication of them, and no selection against extending it still further (**4.1** and **4.2**).

To sum up: although we cannot be certain that no animals are conscious, we can say that it is most unlikely that any of them are. Science does not deal in certainties but in order to keep going it must adopt working hypotheses, the most plausible at the time. These are by common consent treated as 'true' until replaced by more plausible ones. It is in that spirit that anthropomorphism is treated here as a definite mistake. In point of fact, the hypothesis that animals are conscious is not a scientific one, since it cannot be tested.

Once arrived at, that verdict is promptly confronted by the entrenched, age-old, anthropomorphic belief that animals are conscious and purposeful. This seems to be one of the things built into human beings by both nature and nurture so that we cannot hope it will simply fade away. Many neobehaviourists will not be as ready as I am to take an uncompromisingly negative view of animal consciousness but the case against anthropomorphism does not depend on that view. Readers' unwillingness to accept that argument does not mean that they are prepared to go to the other extreme and *assume* that animals are conscious. My main purpose here is not to persuade people that animals are unconscious but rather to bring out the danger of unthinkingly assuming that they *are* conscious. All of us without exception, including those who are convinced intellectually that anthropomorphism is a mistake, remain in danger of falling into it without noticing (e.g. p. 8). Clark Hull was a radical behaviourist and as such is out of favour nowadays, but that does not invalidate his warning: "Even when fully aware of anthropomorphic subjectivism and its dangers, the most careful and experienced thinker is likely to find himself a victim to its seduction. Indeed, despite the most conscientious effort to avoid this it is altogether probable that there may be found in various parts of the present work hidden elements of the anthropomorphically subjective" (Hull 1943, p. 27). Admittedly that is a danger that most neobehaviourists no longer take seriously. They consider that the battle against anthropomorphism was won and has passed into history. The following Chapters 3 to 6 present grounds for questioning that view and it will be returned to in Chapter 7.

CHAPTER 3

3.1 Instinct

The history of ethology must be almost unique. Its founders were awarded a Nobel Prize, which they richly deserved, and yet the theoretical core of the discipline as they founded it survived only a few decades. Most of the overhauling occurred in the 1950s and took the form not so much of revising that original theoretical core, the theory of instinct, as of simply demolishing it, leaving nothing of comparable generality in its place. By 1968 the theory of instinct put forward by Lorenz (1937, 1950) and Tinbergen (1950, 1951) and commended by Thorpe (1948, 1954), could be alluded to by Bateson, then a seasoned editor of *Animal Behaviour*, in the following terms:

> "Worship of the old gods and the intellectual baggage that went with it still survives quaintly in odd corners. But for the most part proponents of a Grand Theory have either been forced to close their eyes to awkward evidence or modify their ideas to the point of unfalsifiability. Explanations have thus become more limited in scope. (Bateson 1968)

This was not an unfair judgement. It has been echoed by R. Dawkins (1976*a*) and Barlow (1989) and repeated recently by Bateson & Klopfer (1989). Ridley (1982) even described a book in which Lorenz (1981) defended his original ideas, as "awkwardly antediluvian".

The 'Grand Theory' held that each major behavioural system was driven from within by outward-flowing nervous "energy" (Lorenz) or "motivational impulses" (Tinbergen) which were specific for that behaviour and accumulated until released, usually but not necessarily by an external stimulus. Since the causal drive was supposed to be specific to the behaviour that it produced, this concept of drive is equivalent to what Tinbergen (1942) had earlier called a "linear causal chain", an idea he rejected at that time. This in turn was equivalent to the term "unitary drive", which Hinde (1959*b*) introduced subsequently and likewise rejected as an idea. Also rejected by the critics was the over emphasis on internal causal factors, with non-recognition of sensory feedback and reduction of the role of external stimuli to that of merely blocking or releasing of energy flow, and the lack of any role for inhibition despite its known great importance physiologically. The concepts of behavioural causation in the Grand Theory were criticized briefly by Kennedy (1954) and extensively by Hinde 1959*a*, *b*, 1960), and the amendments were in due course adopted by Tinbergen (1963, 1969) if not by Lorenz. Lehrman's (1953) criticisms were aimed mainly at the supposed innateness of instinct: which is not the main concern of this book. Now that the dust has long settled M. S. Dawkins (1986), Colgan (1989) and Manning (1989) have paid tribute to the faded grandeur of the Lorenz–Tinbergen theory of instinctive behaviour but at the same time set forth the continuing strong objections to it. M. S. Dawkins summed up its present status as follows:

> "Most contemporary textbooks on animal behaviour tend to dismiss 'instinct' altogether and attempt to consign it to honourable retirement, together with 'fixed action patterns', 'vacuum' and 'displacement activities', 'releasers' and several other derived concepts once in common use.' (M. S. Dawkins 1986, p. 67)

How did it come about that such ideas, soon to be jettisoned as mistaken, were so readily accepted in the first place? This would seem to be a very natural question but it was not raised by any other critics mentioned in the paragraphs above. The idea of behaviour being driven entirely from within could be explained, in part, as an overreaction to the radical behaviourists' minimizing of the role of internal factors. But that was not all. The early ethologists were mostly people of a naturalist bent who had not been inoculated against anthropomorphism by the radical behaviourists. It is therefore not surprising that Tinbergen found reason to remark on its occurrence among those ethologists: "There has been, and still is, a certain tendency to answer the causal question by merely pointing to the goal, end or purpose of behaviour.... This tendency is... seriously hampering the progress of ethology" (Tinbergen 1951, p. 4). Tinbergen himself then put his finger on the source of that teleological tendency, as already quoted in part on p. 28:

> "The main reason... this type of deviation from the natural course of causal study has such a tenacious hold, especially in animal ethology, is that introspection leads us to believe that our own behaviour is controlled, to a certain extent, by 'foreknowledge' of ends or goals."

Taking a human mental experience and imputing it to animals in that way is of course anthropomorphism, albeit unconscious. It is an example of our built-in tendency to think of animal behaviour subjectively as we think of our own. Furthermore, the Grand Theory resembled Freud's already-established, wholly subjective theory of human behaviour in so many ways (Carthy 1951; Kennedy 1954) that one must suspect that this was a second source of the unwitting anthropomorphism of the early ethologists, although there is no published evidence of that. But most critics of the Grand Theory did not even mention the likelihood that anthropomorphism lay at the root of the mistakes that they pointed out.

Like the Grand Theory of instinct, the six further cases of misconceptions traceable to anthropomorphism that are to be discussed in this chapter are cases which are generally recognized already to be errors of one kind or another.

3.2 Nest-building

Thorpe (1956, 1963) interpreted nest-building by birds as purposive or 'goal-directed' in his sense of "striving after a future goal retained as some kind of image or idea" (see Chapter 1). This was a simple and clear-cut instance of a mistake due to the teleological thinking that Tinbergen deplored. Crook (1960, 1965) carried out detailed studies of several species of African and Indian weaverbirds building their elaborate basketwork nests. He had Thorpe's interpretation in mind and summed it up thus: "Construction appears as if done to a 'plan', deviation from which is corrected by some process of comparison of the actual structure with the required 'ideal' completed nest." But his own observations, Crook concluded, did not "support the idea of a process of comparison with a nest plan, as it were, in the bird's mind". M. S. Dawkins (1983) referred to studies of nest-building by a variety of birds and commented that: "Some male weaverbirds are polygamous and may have several nests under construction all at once, so that any simple 'fixed sequence' of building can be ruled out straight away.... The ability to repair damaged nests might suggest that the birds have an image of what a completed nest should look like. Collias & Collias, however, pointed out that the actual movements that the bird makes in both construction and repair "...are very stereotyped...". If it stands in the right place and carried out its stereotyped building movements, the result is a repaired nest, but there is no necessary implication that the bird is comparing the nest's present structure with some kind of ideal pattern" (M. S. Dawkins 1983, pp. 93 and 94).

In their wide-ranging discussion of "Goals and response control", Hinde & Stevenson (1970) had been no less sceptical about Thorpe's purposive interpretation of nest-building. It is again noteworthy that none of the authors suggested why Thorpe's idea could not be confirmed, although the reason is plain: it was an anthropomorphic idea. The assumption that behaviour sequences are guided by comparison of the current sensory input with a mental image of the goal has been recognized as misleading in this particular case and the next (see **3.3**), but it is still very much alive, as will be seen in Chapter 4.

3.3 Search images

A better-known variant of Thorpe's teleological idea of a goal image serving to guide behaviour is the notion that bird predators form "search images" of cryptic prey. It was supposed that the predator rapidly "learned to see" the camouflaged prey as a result of finding one by chance, and so captured more of them from then on. This concept entered ethological thinking relatively late when the original Grand Theory of the founder ethologists (see **3.1**) had already come under attack, but this particular concept has been called into question only recently.

The term 'search image' was coined by von Uexkull (1934), referring to our own use of a mental image of a lost object when searching for it. Some considerable time later, de Ruiter's (1952) laboratory observations of jays and chaffinches foraging among cryptic stick-insect larvae were seen as demonstrating the occurrence of a similar phenomenon in birds. But the idea became widely known and accepted later still, when L. Tinbergen (1960) invoked it in interpreting the results of his eight-year study of fluctuations in the species composition of the insects eaten by great tits in Dutch pinewoods. He attributed these dietary shifts to the birds adopting a *specific search image* for each prey species and therefore coming to specialize on it once it had become numerous enough for a specimen to be encountered by chance.

Royama (1970) accepted the possibility of search image formation but could not confirm Tinbergen's conclusions and did not agree that they followed from the data (but see Curio 1976). He explained that he had criticized Tinbergen extensively because "his theory has become so popular among ornithologists and ecologists in recent years that it is regarded almost as fact: the hypothesis has not been proved and must not be treated as fact until obvious contradictions have been removed". Smith & R. Dawkins (1971) shared this scepticism, and so apparently did Hinde for he seemed to imply, in one passage, that the search image hypothesis was anthropomorphic: "Those who take the view that it is legitimate or even desirable to use human experience to interpret animal behaviour will have no difficulty with the view that an animal can have a search image for stimuli not present at the time. Others, however, will feel happier if they do not extrapolate beyond the observable facts" (Hinde 1970, p. 124). However, when discussing learning in general later in the same book Hinde cited search images without demur, thus:

> "Much appetitive behaviour involves a scanning of the environment until a particular set of stimuli are encountered, and can be understood on the assumption that the animal is using a 'search image'. In other words, searching involves not only a change in behaviour, but also a change in responsiveness not immediately apparent in behaviour." (Hinde 1970, p. 592)

Up until 1987, books, essays and research papers continued in the same vein, treating search image formation as an established fact (Tinbergen 1951; M. S. Dawkins 1971*a*, *b*,

1981, 1986; Murton 1971; Krebs 1973; Curio 1976; Alcock 1979; Pietrewicz & Kamil 1979, 1981; Morse 1980; Wallace 1983; Barnard 1983; Bond 1983; Lawrence & Allen 1983; McFarland 1985; Lawrence 1985*a*, *b*, 1986; Gendron 1986; Ridley 1986; Krebs & Davies 1987: an incomplete list). Curio (1976) voiced the general opinion that "the best evidence for search images to occur and facilitate prey capture has been adduced by M. S. Dawkins (1971*a*, *b*) in her experiments with domestic chicks in which previous experience was accurately controlled". She had presented her chicks with camouflaged and conspicuous rice gains and "after eating one camouflaged grain they rapidly ate more of them.... The chicks learned to see camouflaged grains; they formed a search image" (Ridley 1986).

But later, on reappraising the evidence, Guilford & M. S. Dawkins (1987, 1989*a*, *b*) reported that they could find no evidence in favour of search images after all: "the existence of search images remains not proven". No-one who had claimed to have evidence of search image formation, they argued at length, had eliminated the alternative, and if anything more plausible, hypothesis that the chicks responded to an encounter with cryptic prey simply by spending more time scanning each field of view. This was the "search rate hypothesis" of Gendron (1986; Gendron & Staddon 1983). According to Guilford & M. S. Dawkins (1987), "chicks that had learnt to see cryptic grains [in M. S. Dawkins's (1971*a*, *b*) experiments] simply added this ability to their repertoire of detectable prey. This is just what the search rate hypothesis, but not the search image hypothesis, predicts."

It is remarkable in retrospect how widely the idea of search image formation was accepted once it had been applied to bird predation in nature, and how tenaciously it retained its reputation as an established fact even after the inadequacy of the evidence for it had been pointed out and an alternative explanation of the facts had been proposed. How could that happen? When we ourselves search, of course, we really do have a mental image of what we want to find. The idea of a search image is therefore familiar and instantly acceptable to us without further thought. That seems to be the only reason for the uncritical acceptance of the idea for animals too; although again, this anthropomorphic explanation of the mistake has not hitherto been pointed out.

3.4 Language

Probably the most spectacular examples of professional scientists – animal psychologists, in this case – being misled by their subconscious anthropomorphism come from the last series of American attempts to teach animals our language. In this case the mistake was not due to teleological thinking but to anthropomorphism in the different form of assuming that higher animals share our learning abilities. There was a sustained effort to train dolphins to mimic human speech at the US Navy's Communication Research Institute in Miami in the 1960s. "Numerous rumours and some reports were put in circulation to the effect that the animals ... were indeed capable of reproducing words 'appropriately'" (Sebeok & Umiker-Sebeok 1980, p. 421). Public interest was such that a successful book and film were produced, built round the fiction that scientists could teach dolphins to converse with them. In reality, the scientific publications from the project were found to contain no real evidence of success and it was shut down completely in 1968. "The long shadow of Clever Hans darkened that undertaking from the start ... the whole thing proved to be a great bluff" (ibid., p. 422). The classic "Clever Hans" saga has been fully documented by Umiker-Sebeok & Sebeok (1981). They defined the phenomenon succinctly as "the human observer unintentionally modifying the animal's behaviour to produce the desired results". Note "unintentionally": the human observer may be quite honestly unaware of the signals he is giving, and the animal may be capable of responding discriminately to much smaller movements than the observer imagines. These unexpected difficulties made the unravelling of the Clever Hans case very laborious.

It is a measure of our human eagerness to believe that higher animals possess human-like minds that in the very next decade after the dolphin fiasco much the same story was gone through all over again and on a much larger scale, this time with apes. Twenty-one representative papers from all sides of this highly charged episode are reprinted in Sebeok & Umiker-Sebeok (1980). It is difficult to imagine the excitement, public as well as scientific, over the ape language projects at that time in the U.S.A. Two contemporary comments give something of the flavour. "Today one can scarcely read a newspaper or news magazine without encountering a feature extolling the latest linguistic accomplishments of one or another ape.... With increasing frequency, widely read journals such as *Science* publish reports of the transmutation of base primates into noble ones. It is no wonder there is a growing belief among students and scientists alike that modern behavioural science has in fact succeeded in teaching human language to apes" (Limber 1980). "It is unlikely that any of us will in our lifetimes see again a scientific breakthrough as profound in its implications as the moment when Washoe, the baby chimpanzee [the subject of Gardner & Gardner's (1978) long study], raised her hand and signed 'come-gimme' to a comprehending human" (Hill 1980).

But then the bubble burst, punctured by Terrace *et al.* (1979):

> "The dramatic reports of the Gardners, Premack and Rumbaugh that a chimpanzee could learn substantial vocabularies of words of visual languages and that they

> were also capable of producing [by signs] utterances containing two or more words, raise an obvious fundamental question: are a chimpanzee's multi-word utterances grammatical? In the case of the Gardners, one wants to know whether Washoe's signing *more drink* in order to obtain another cup of juice, or *water bird* upon seeing a swan, were creative juxtapositions of signs. Likewise one wants to know whether Sarah, Premack's main subject, was using a grammatical rule in arranging her plastic chips in the sequence, *Mary give Sarah apple*, and whether Lana, the subject of a related study by Rumbaugh, exhibited knowledge of a grammatical rule in producing the sequence *please machine give apple*.... One has too little information [on the order in which signs were 'uttered'] to judge whether such utterances are manifestations of a simple grammatical rule or whether they are merely sequences of contextually related signs...." (Terrace 1984*a*, pp. 179–80 and 182)

At first Terrace (1979) had believed that his records showed that his own chimpanzee subject, Nim, did use some simple grammatical rules, in agreement with the other workers' claims. But he began to have doubts when he went through the videotapes and was struck by how much more the human trainers had influenced Nim's signing than his trainers had realized at the time. While a training session was in progress their attention and recording was concentrated on Nim's behaviour, not on what Nim for his part was watching, namely, their own efforts to encourage his signing and their own reactions to his performance. Terrace then managed to obtain from several other laboratories films or videotapes of some at least of their working sessions, in order to analyse not only the animal's signing actions but also the trainer's words and actions to which the animal was responding. With hindsight, such "discourse analyses" were an obvious requirement in a study of communication but they had not previously been done with the chimpanzees. From these analyses Terrace concluded that "Nim's, Washoe's, Ally's, Booee's [*all chimpanzees*] and Koko's [*a gorilla*] use of signs suggests a type of interaction between an ape and its trainer that has little to do with human language. In each instance the sole function of the ape's signing appears to be to request various rewards that can be obtained only by signing. Little, if any, evidence is available that an ape signs in order to exchange information with its trainer, as opposed to simply demanding some object or activity" (Terrace 1984*a*, p. 197).

Savage-Rumbaugh *et al.* (1980) accepted and even enlarged upon Terrace's main criticism, and revised their research objectives (Savage-Rumbaugh 1986). "Two factors" responsible for this debacle "were unrealistically ambitious goals and the uncritical acceptance of the purported symbols used by the apes" (Terrace 1986, p. xiv). The goals were overambitious in that the researchers were looking to the apes for nothing less than grammatical human language, sentence-building. They "focused almost exclusively on evidence of grammatical competence" (ibid.), and the reason they did so was "the psycholinguist's emphasis on grammatical competence as the hallmark of human language" (ibid.). The researchers were preoccupied, not with how the apes used symbolic communication, but with getting them to use it in the way that we do. In short, their judgement was distorted by anthropomorphism – although Terrace did not identify that as the root of the trouble.

The anthropomorphism did serious harm. Not only did it lead to uncritical acceptance of inadequate evidence for what the researchers wanted to show, but it meant they "missed a wealth of important issues about symbol use" by the apes (Terrace 1986, p. xvi; cp. Seyfarth 1982, p. 402). Not surprisingly, the apes' use of symbols is much simpler than ours (although subsequently Savage-Rumbaugh (1986) showed that it was not as simple as Terrace had thought). The damage did not stop there. When the bubble of sentence-building burst "scepticism about an ape's grammatical competence had the unfortunate effect of stifling interest in other significant questions about linguistic competence" (Terrace 1986, p. xvi), and "the widespread interest that this research program generated was followed quickly by a reaction that has bordered on benign neglect" (ibid., p. xiv; see also Ristau & Robbins 1982).

It was not only the "very ancient... intense desire to enter into direct dialogue with animals" (Hediger 1980) that led many people, including qualified scientists, into believing that apes could be taught to use visual signals to communicate in grammatical human language. There was a great deal of mutual criticism among the several competing teams, to the effect that the others were overinterpreting their results. Savage-Rambaugh *et al.* (1978) declared "we believe that there is no evidence, other than richly-interpreted anecdote, to suggest that Washoe and other signing apes are producing anything more than short-circuited iconic sequences." Gardner & Gardner (1978) for their part declared "the results that Premack and Rumbaugh et al. have presented thus far are more parsimoniously interpreted in terms of such classic factors as Clever Hans cues, rotememory, and learning sets". According to Hediger this was a pot-and-kettle situation. Both sides were right, for the Clever Hans Fallacy had ensnared all the parties. He had studied the projects and decided that none of the researchers was sufficiently alert to the danger of Clever Hans illusions. It was a "major fallacy underlying all such work that scientists, even those unfamiliar with apes, can be counted on to detect social cues and/or other forms of self-deceptive manipulative behavior at play in the interactions between man and ape" (Hediger 1980). In all the projects "the experimental animal is supposed to learn signs in order to use them in a way that makes sense, that is, the sense the leader of the experiment desires. Experience has shown it always to be extremely difficult to decide if the animal actually understands and uses the signs that are presented to it, or if it reacts to involuntary signs (like expressions or the movements of the leader of the experiment)..." (ibid.).

Hediger was writing at a time when the Lana Project (Rumbaugh 1977), which was claimed to reach the highest degree of objectivity and non-influence over the animal, was still in progress; and he went on: "from the beginning, the members of the team involved with the Lana Million Dollar Project... emphatically hoped for the success of their... project. This is a basically dangerous starting point. We have to ask ourselves very seriously if the Lana Project will reach the same fate as the alleged dialogue with dolphins" (Hediger 1980). It did (Thompson & Church 1980).

The danger of Clever Hans illusions will be the greater, the more empathy the experimenters feel for their subjects and the more determined they are to succeed in their efforts to 'humanize' them. Both factors are particularly powerful for workers on higher primates, chimpanzees especially. Indeed it is common for close students of these animals to think of them anthropomorphically (see **2.6**). Ann Premack, who collaborated in the attempt to teach human language to the chimpanzee Sarah, related how "people who raise chimps have high expectations of them as they have for their own children, and when the chimps don't perform at these levels, the 'parents' are often bitter" (quoted by Sebeok & Umiker-Sebeok 1980, p. 5). With such a will to win on the part of the trainers, it is hardly surprising that their endeavours turned out to have been dogged by unconscious anthropomorphism, mostly in the shape of Clever Hans delusions. The fact that better evidence of grammatical ability, notably in *Pan paniscus* (Greenfield &

Savage-Rumbaugh 1990), has since been forthcoming does not erase the lesson to be learnt from the earlier anthropomorphic errors of interpretation.

3.5 Imitation

The evidence that has been widely supposed to show that one animal can learn to reproduce virtually the exact behaviour of another, that is truly to imitate it, was succinctly reviewed by Bonner (1980) and Davis (1981) and earlier papers on the subject were reprinted by Klopfer & Hailman (1972). This turns out to be another case of too-readily assuming that animals share all our learning abilities. Imitation, a remarkable feat considering the multi-staged neural processing that it presumably requires, is a major component of behavioural development in children from an extremely early age. As for animals, the best-known and probably the first scientifically recorded case of supposed imitation is the wide and fast spread of the habit of tearing open milk bottles by birds, mainly blue tits, in Britain. Imitation has long been seen as a component of cultural inheritance in sub-human primates also, the most famous case being the spread of a food-washing habit in successive generations of a colony of Japanese macaques (Miyadi 1964; Itani & Nishimura 1973). Much in the same way as it was for long assumed that birds were using specific 'search images' in finding cryptic prey although the possibility that they were simply 'searching' more slowly had not been ruled out (see **3.3**), so it transpires that it has often been assumed that a new habit is spread by imitation, although a simpler possibility had not been ruled out. Thorpe (1951, 1963) originally described the simpler possibility as "local enhancement". Whiten (1989) has recently redefined this under the more indicative name of "stimulus enhancement", when "a performer's action merely focuses an observer's attention on critical environmental features, increasing the speed with which the observer subsequently learns a similar behaviour pattern through its own efforts".

The original account of the spread of milk bottle opening by tits by Fisher & Hinde (1949) concluded by asking "How far did the individual birds learn the habit from each other, or invent it for themselves?" While insisting that a proper answer could be obtained "only from carefully controlled experiments on birds of known history" they mentioned one possibility which was in effect "stimulus enhancement" by pioneer birds who exposed the milk and thus encouraged others to learn to open bottles for themselves. They suggested "that those birds that first drank from milk bottles without having previously seen others do so, drank from bottles which had already been opened, or from bottles in which the top was awry. One or two occasions of this type may have been enough to produce an association between the milk bottle and food." On the other hand, they argued, "if it is assumed that the first occurrences of the habit were not dependent on some accident such as a misplaced bottle top, then it would seem that in each district the milk bottles presented equal sensory clues to all birds which came near them, that only certain rather exceptional birds were able to profit by them, and that the subsequent learning of the habit by other individuals in the same district took place by some sort of imitation". The authors (ibid.) decided that the circumstantial evidence clearly favoured that second hypothesis, "that the practice has been begun by comparatively few birds and that the vast majority of tits have learned it in some way from others".

In a second paper still quoting field records only but now acknowledging advice from W. H. Thorpe and L. Tinbergen, Hinde & Fisher (1951) switched their preference to 'local enhancement' thereby implicitly rejecting the imitation alternative. Thorpe (1963) believed that imitation was restricted to primates and bird song. Some years later, however, Hinde & Fisher (1972) said that their earlier papers had displayed "some of the worst faults of ethology at that time, such as the tendency to regard descriptive labels as explanatory," and rejected local enhancement on the grounds that it had little explanatory value. They did not explicitly embrace the hypothesis of imitation but mentioned no other alternative to enhancement. Since then both the bird and milk bottle story and the monkey food-washing story have been generally accepted as illustrating imitation (e.g. by Klopfer 1961; Itani & Nishimura 1973; Wilson 1975; Bonner 1980; Davis 1981; Kummer & Goodall 1985).

Recently, however, scepticism has set in (Visalberghi & Fragaszy 1990*b*; Galef 1990). Whiten (1989) has summarized a number of new primate studies that favour 'stimulus enhancement' and quotes the conclusion of some recent workers on chimpanzees: "Reports of apes imitating ... have not systematically controlled for other forms of observational learning .... We are aware of no evidence that any species other than human beings is capable of copying the precise topography of a conspecific's sensorimotor task" (Tomasello *et al.* cited by Whiten 1989). In accord with that, Sherry & Galef (1984, 1990), for example, have demonstrated experimentally that naive Canadian tits do not learn to open milk bottles when allowed to watch others doing so, that is by imitation, but do learn to do it by feeding from milk bottles opened by others (stimulus enhancement), and also, rather surprisingly, they learn to do it for themselves provided they can see a conspecific, even if that bird has no bottle and cannot give any demonstration of bottle-opening.

No doubt has yet been cast on the ability of some birds to learn some songs by imitation, and there is one piece of experimental work apparently showing that pigeons can learn to open food containers by imitation (Palameta & Lefebvre 1985). But there has clearly been an undue readiness to assume that learning has occurred by imitation without the alternative of stimulus enhancement having been ruled out. The only apparent reason for that unfounded assumption that animals were "capable of copying the precise topography of a conspecific's sensorimotor task" is our familiarity with this capability in ourselves or, in other words, once again, the influence of anthropomorphism – although this has seldom been said (but see Visalberghi & Fragaszy 1990*a*).

3.6 Functions as causes

Mistaking the function of a piece of behaviour for its cause – a special case of reversing cause and effect – underlies many teleological errors. The difference between causal and functional descriptions of behaviour has been generally recognized since Tinbergen's (1951) original statement but it has not been upheld consistently by any means. (Note that some purely terminological confusion arises from the fact that functional descriptions are also causal, although in a different sense, see below.) If, in the past, a piece of behaviour had the effect of increasing the (inclusive) fitness of the individual, then natural selection will have favoured the perpetuation of that piece of behaviour together with the machinery that produces it; otherwise they would not still be with us. Its function is therefore called the evolutionary or *ultimate* cause of any behaviour we see today, while the causal mechanism of the same behaviour as it works today is called the *proximate* cause of that behaviour (a difference lucidly expounded by Alcock (1979, pp. 2–7). The *mechanism* or machinery of the behaviour refers to the events occurring at lower levels of physiological integration that generate the behaviour. Studies of these two kinds of cause of one piece of behaviour are undoubtedly complementary but it is equally important to remember that ultimate and proximate causes are quite different things. Confusing them leads to mistakes.

> "Why questions require an approach that is in some ways totally different from that used by researchers intrigued by proximate questions." (Alcock 1979, p. 7)
> "We must be clear whether we mean a causal or functional category when we use terms like 'sexual' or 'aggressive' behaviour." (Hinde 1982, p. 32)
> "Causal analysis does not indicate adaptiveness, and functional analysis does not reveal causal mechanism." (Colgan 1989, p. 133)
> "It is not good practice to say that an animal adopts a cryptic posture in order to avoid predators, since this implies that the evasion of predators is a motivational cause of the behaviour, whereas it is really a function of the behaviour." (McFarland 1981)
> "A great deal of misunderstanding can arise if functional and causal ('How?') explanations are confused." (Krebs & Davies 1987, p. 351)

In fact, they are often confused; otherwise these warnings would be superfluous. "Factors influencing survival value are sometimes called 'ultimate' while [immediate] causal factors are referred to as 'proximate'. It is these two answers [to the causal question] that are the most frequently muddled up..." (Krebs & Davies 1981, p. 5). "Mechanisms and purposes have been confused very often" (Ridley 1986, p. 7). The confusion is exemplified in this passage: "The determinants of the initiation and termination of feeding bouts are not changes in internal states but are ecological variables, such as cost, abundance, availability, and value.... The effect of deprivation...most likely reflects the animal's estimates of changing abundance and availability. Finally, it seems likely that what we have called *motivation* represents the animal's attempt to solve the problem of optimal allocation of time and energy among biologically significant activities" (Collier 1980, p. 148). There the writer first presents some ultimate causes, namely ecological variables, as if they embodied sensory stimuli that start and stop feeding today; that is to say as if they were proximate causes. Then in the final sentence he confuses a proximate cause, namely the animal's motivation in attempting to do something, with an ultimate cause, namely natural selection optimizing the animal's allocation of time and energy.

It is important to note that the confusion of proximate and ultimate causes does not work equally in both directions, a qualification that is rarely made explicit. More exactly, what should be said is that ultimate causes are often mistaken for proximate ones, and not vice versa. This is because the confusion usually arises from anthropomorphism which has the specific effect of making ultimate causes appear proximate. The function or ultimate cause of an action is the end-result of that action. When that end is regarded anthropomorphically as the purpose of the action then the end is seen as an antecedent, and therefore proximate, cause of it; that is teleology, a reversal of cause and effect. This cart-before-the-horse kind of reasoning was neatly satirized by Lorenz (1950): "The next time a child asks me what makes a train go, I shall answer that this is caused by a special factor, called the locomotive force." Or, as Tinbergen put it, more literally: "When the conclusion that the animal hunts because it is hungry is taken literally, as a *causal explanation*, and when it is claimed that the subjective phenomenon of hunger is one of the causes of food-seeking behaviour, physiological and psychological thinking are confused" (Tinbergen 1951, pp. 4–5).

To take another example, the two kinds of cause are confused when an animal is described as 'searching'. We find this a very satisfying description of what it is doing – all too satisfying, for it arouses no curiosity about the behaviour itself. It fails to say what exactly the animal is doing. We can guess it is making some kind of scanning movements, but these might be movements of the whole animal (locomotion) or movements only of the parts of its body bearing sense organs, like movable eyes, or antennae. Ignoring that, we tend to accept mere 'searching' as a description instantly, unthinkingly, because we know so well what it feels like to search. "In every day language the word searching characterizes a certain state of mind rather than a series of movements" (Tinbergen 1942, p. 59; cp. R. Dawkins 1976*b*, quoted on p. 25, above). What the animal is "up to" here seems abundantly clear to us. That is what we are always most curious about and most satisfied to know: the function of the behaviour. So we do not feel as curious about the animal's actual movements, or the external stimulus situation, as we otherwise should, and we feel no pressing need to look at them more closely. To describe the animal's behaviour by a subjective term, 'searching', is to imply a purpose behind the behaviour, which as Tinbergen said seems like a (proximately) causal explanation. Using such a term can therefore easily create the false impression that the animal's behaviour has been explained not only functionally, but also proximately.

Again, Krebs & Davies identified a criticism commonly levelled at their book *Behavioural ecology* as an instance of an ultimate cause being mistaken for a proximate one: "an objection that is often raised to labels such as 'selfish', 'spiteful', 'sneak', 'transvestite', and 'rape' used by behavioural ecologists ... is that the labels are too anthropomorphic and are loaded with the implication that animals are endowed with human-like motives. The answer to this objection is that the labels are used not to describe the causal mechanisms underlying the behaviour but to describe its functional consequences"

(Krebs & Davies 1987, p. 351). That makes it quite clear that the authors themselves are not for a moment intending their anthropomorphic language to be taken literally. They are neobehaviourists, and their own use of such language is wholly metaphorical, what will here be called 'mock anthropomorphism' (p. 88). But they are overconfident that merely saying they are not being anthropomorphic will prevent other people from being misled by the subjective overtones of the language. Indeed the authors themselves implicitly recognize this danger with their warnings against confusing proximate and ultimate causes quoted above. As an illustration, Estep & Bruce cited the indiscriminate application of the term 'rape' to a variety of animals by various authors, as evidence for a conclusion close to the viewpoint of this book:

> "to apply a human label to the behaviour of non-humans does not necessarily make the events the same. Indeed, to use such a label may imply false similarities and mislead about the motivation (proximal causes) and function (ultimate causes) of the animal's behaviour.... Rape is an emotionally charged word that carries with it a wide range of social and ethical implications. By using the term to describe non-human behaviour, we are forcing certain human cultural standards on non-humans. We assume that *scientists who apply the term to non-human behaviour do not intend these connotations, yet they cannot be avoided.*" (Estep & Bruce 1981, p. 1273; my emphasis)

Dunbar (quoted on p. 26) and Baker (see **4.1**) illustrate how even scientists can take the function, that is the ultimate cause, of a given behaviour as its proximate cause. It could be said this is not very surprising since Dunbar and Baker are explicit anthropomorphists. More important, neobehaviourists sometimes fall into the same error unintentionally, as we have seen in a number of instances, despite warnings against it from

Tinbergen (1951, 1963) and others cited earlier in this section and in **4.2**. Probably the most frequent cases of mistaking ultimate causes for proximate ones occur in the labelling of motivations, to which Hinde (1982) alludes in the quotation on p. 50. In his earlier book Hinde (1970, p. 417) had already noted that 'drives' are "labelled in terms of their functional consequences", and such teleological labelling is still common practice. Unfortunately Manning's (1979) particularly useful and highly successful textbook set the example when he said that the *motivation* of a piece of animal behaviour is to be identified by the *function* of that behaviour:

> "we usually have to deduce how an animal is motivated by observing how it behaves... we have to identify motivation with reference to a function that seems reasonable to us. If one animal attacks another we ascribe this to aggressive motivation; if it eats we ascribe this to feeding motivation and so on." (Manning 1979, p. 109)

As a specific illustration of Lorenz's satirical parable of the teleological "locomotive force", above, this could hardly be bettered! It is almost as 'good' as my own equivalent, the migratory locusts' "locomotory drive" cited on p. 8. Regrettably, Manning was quite right in his frank statement that the usual way of judging what is the 'motivation' of any given behaviour is by referring to its function, notwithstanding Tinbergen's warning against this (pp. 35 and 52, above). This is the crux of the matter; such teleological errors may seem astonishing but they are hard to avoid because they are made under the pressure of what McFarland (1989*a*, 1989*c*) has dubbed the "Teleological Imperative" in the heads of us all.

those who study mechanisms, there is a special sense in which Baker (1982; see **3.1**) is right that the former group now dominates the scene – although not in his sense. It is more difficult than ever for the students of mechanism to resist using vivid anthropomorphic language and thus staying in the swim alongside the students of function, even when they know that using this language risks misconceiving the mechanisms of behaviour. Dunbar and Baker both attribute their explicit anthropomorphism to the growing influence of sociobiology and behavioural ecology where the emphasis is on 'strategies' and 'decision-making' which they choose to interpret as mental, that is proximate, causes. Of course they really have no justification for making that attribution, since as we saw on p. 53 neobehaviourists do not necessarily interpret 'decision-making' as mental. All the same, the attribution is no accident. Sociobiology and behavioural ecology are subjects in which anthropomorphic language is defensible as 'mock' anthropomorphism (see **5.1**), and their expansion has been at the expense of the study of proximate mechanisms. The routine use of anthropomorphic language, being now overwhelmingly fashionable, has spread into the study of proximate causes too, where such language (see **4.1**) is on the contrary indefensible anthropomorphism.

Moreover, insofar as behavioural ecology's concentration on analysing the ultimate causes of behaviour is at the expense of analysis of the proximate causes, it is also at the expense of cross-fertilization between these two endeavours. That can be very fruitful as Krebs & Davies (1981), Toates & Birke (1982), M. S. Dawkins (1989) and Bateson & Klopfer (1989) aver. Most important, the lop-sided progress of ethology tends to (but see p. 156) defer indefinitely the desirable bridging of the causal "gap between what nerve cells do and how animals behave" in Roeder's (1965) phrase. Bullock (1965) reckoned that this gap was "probably wider than any other gap between levels of integration in natural science", and it still is. The high hopes of the 1950s that the gap would be closed in the foreseeable future have been disappointed: "the optimism of the fifties has not been fulfilled" (Hinde 1982, pp. 170–1; echoed by Halliday & Slater 1983; Ridley 1986; Manning 1989).

M. S. Dawkins (1989) places the major responsibility for the present imbalance firmly on the ethological side of the gap "where we are just not doing our bit". No doubt that is a major reason for the gap remaining wide. Four books in the new field of 'neuroethology' (Ewert 1980; Guthrie 1980; Camhi 1984; Young 1989) bear witness to the way the physiologists have been doing their bit on their side of the gap. Admittedly, neuroethology has not yet lived up to the full promise of its challenging name. Ethology itself deals of course with the integrated behaviour of the whole animal with its full behavioural repertoire 'in play', whereas the triumphs of neuroethology so far have come from the neurophysiological analysis of a single behaviour system such as prey-catching. But, that aside, it seems safe to say the neurophysiologists would have made much more progress if they had not been so hampered by the "distinctly old-fashioned" (M. S. Dawkins 1989) ideas about behaviour that they got from ethology when it was young and the Lorenz–Tinbergen Grand Theory of instinct (see **3.1**) was in the ascendant. Those ideas caught the imagination of certain influential animal neurophysiologists such as Bullock (1965) and Hoyle (1965) and held it even after ethologists had come to see the ideas as erroneous (M. S. Dawkins 1986; also **3.1**). Neuroethologists trying to work with ethological concepts and terms still find them unhelpfully vague. "Neurobiological descriptions are often precise and detailed, and unless they can be matched by similar sharpness of meaning in behavioural observations, it is difficult to make correlations between them. For the moment, it is necessary to accept, and if possible to adapt the terminology developed by ethologists, and see how it corresponds to the kinds of problem that neurobiologists are equipped to study" (Guthrie 1980, p. 18).

The neurophysiologists may not have realized that their difficulties stemmed from the psychological origin and overtones of the concepts and terms of early ethology, but these were simply not concepts that they could use. The psychological concepts were physiologically too intangible to offer problems that neurophysiologists could hope to tackle experimentally at their level of analysis (Tinbergen 1969; Staddon 1983; Kennedy 1987*a*; M. S. Dawkins 1989). Hinde (1970, p. 8) had advised the student "to aim at an analysis of behaviour which the physiologist will be able to handle", but M. S. Dawkins (1989) admitted that ethologists have done little to oblige the neurophysiologists in that way. The latter have picked their own behavioural problems and made their own behavioural analyses before getting down to neurophysiological analysis, and for that they could not be blamed. Nor could the physiologists be blamed entirely for "the *naïveté* of premature searches for parallels between the products of ethological analysis and phenomena at the neuronal level" (Hinde 1982, p. 171). Ethologists have themselves made it more difficult to close Roeder's great gap in our knowledge "between what nerve cells do and how animals behave' by grossly underestimating the distance between those two integrative levels in higher animals (Tinbergen 1954, cited on p. 126).

CHAPTER 4

The eleven subject-sections that make up Chapters 4 to 6, discuss notions that are more controversial than those in the preceding chapter. There is as yet no general agreement among neobehaviourists that these notions are erroneous, but neither are they entirely accepted and it will be argued that they are in fact erroneous. All the case-studies in this chapter concern errors arising from teleological thinking, the most common form of unwitting anthropomorphism. Three of the essays in Chapter 3, comprising **3.2**, **3.3** and **3.5**, came into that category, wherein an animal's behaviour is supposedly 'goal-directed', meaning that the animal attains a certain end situation by making compensatory responses to any discrepancy between its current sensory input and some sort of internal representation or 'mental image' of that 'goal' situation. This chapter presents four more cases in the same category.

4.1 Migration

Baker's (1978) monumental treatise on migration is influential because it has no competitor, being the only fairly recent single-author survey of the whole field of migration and an invaluable mine of references. He used the same data in condensed form in a subsequent book (Baker 1982) as a vehicle for fuller exposition of his entirely new view of migration, which he says (ibid., p. 3) required a "major purge of the old ideas" as part of the "behavioural ecology revolution" whereby behavioural ecology, being "a new ideology, completely different from ethology" (ibid., p. v), became "the new establishment" in the study of animal behaviour (ibid., p. 1). He claimed that the behaviour of a migrating animal is controlled by a mental image of a goal on each leg of its travels.

It cannot be said that this idea has been generally recognized as erroneous because Baker's confidently iconoclastic books have had a mixed reception. Some authors have accepted them at face value (e.g. Davies 1981; Barnard 1983; Ridley 1986). Others, who might have been expected to discuss them, mention them only in passing without comment (e.g. Dingle 1980; Able 1980; Huntingford 1984) or do not mention them at all (e.g. Bonner 1980; Krebs & Davies 1984, 1987; McFarland 1985; Slater 1985; Roughgarden *et al.* 1989). Some authors may have refrained from discussing Baker on migration because they had the same difficulties as reviewers of his two books on migration in *Animal Behaviour*. "Four years ago a reviewer (Evans 1980) of Robin Baker's (1978) previous massive tome, *The evolutionary ecology of animal migration*, hoped that the author would 'attempt a much shorter book, written in more straightforward language and with more carefully chosen examples that test his ideas critically.... If he does not, I fear that few readers will have the patience to stay with the present volume.'...Once again he [Baker 1982] presents the core hypothesis of his previous work, namely his 'least-navigation-familiar-area map'; this is of such a general nature as to be virtually useless...the opaque nature of his writing makes his views and ideas obscure" (Swingland 1984).

These reviewers did not say that anthropomorphism lay at the root of the obscurity of the ideas in these books. But in the Preface and first chapter of his 1982 book Baker himself made it clear that anthropomorphism was his basic tenet and he harked back to it throughout: "animals explored, made judgements, and in general solved their spatial and temporal problems in much the same way as Man" (Baker 1982, p. v). An animal was "a sentient organism" (p. vi). It was on that explicitly anthropomorphic premise that he went on to categorize the migratory movements of vertebrates and some invertebrates as moves that are "calculated" by the animals using the "cerebral sense of location" that they have acquired by prior exploratory activity. By contrast, the movements of many invertebrates he considered were based only on "a sense of direction" (if that), and are therefore classed as "non-calculated" (Baker 1982, p. 30). In both groups a mental image of the goal informed every movement: "the individual is forever aware of how things should feel and spends its life chasing a position in its environment in which everything will feel as it should.... If the best of an animal's feeding sites, say, do not match up to its innate mental image of how such a site should be then it begins to search for such a site" (Baker 1982, pp. 203–4).

Baker's conviction that animals feel and think as we do seems stronger than the reasoning behind it. He began with a truism: "there are no grounds for believing that Man is qualitatively unique in the way that an individual's past experience interacts with genetic predisposition to produce observed behaviour". Then, a non-sequitur: "in the absence of such grounds, the same approach and yardstick has to be applied to all animals", so that "if Man is a sentient creature, we have to allow other animals also to be sentient. If other animals are not sentient, then neither is Man" (Baker 1982, p. 2). He argued that this view is part of "the behavioural ecology revolution" (ibid., p. 1). The notion of a revolution stemmed from his belief that "behavioural ecology has taken the place of animal psychology and ethology as the new establishment for behavioural theorists" (ibid.), but what he called ethology there is in fact radical behaviourism. What took the place of radical behaviourism was ethology for zoologists and 'methodological behaviorism' for psychologists. Behavioural ecology came along later. Ethology sees animals, Baker says, "only as bundles of innate reflexes travelling from one automatic response to another" (ibid., pp. v–vi). "Before the behavioural ecology revolution an animal was seen as an automaton, a very simple machine" (ibid., p. 1), and migrant animals were "programmed to respond in a stereotyped, inflexible way to sequences of environmental stimuli" (ibid., p. 13). This is a travesty of ethology. It is common knowledge that the historic achievement of ethologists was on the contrary to discredit such simplistic ideas of the radical behaviourists by bringing out the variability of behaviour and therefore the role of internal causes. Moreover, most behavioural ecologists are not anthropomorphists (see M. S. Dawkins (1986) or the passage on p. 351 of Krebs & Davies (1987) quoted on p. 52, above).

However, Baker's mistaken idea of ethology means that his anthropomorphic view of animals comes across to the reader as the only alternative to a crudely reductionist one such as nobody would dream of adopting nowadays. Again and again, Baker mentions those alternatives only. But a neobehaviourist ethologist has a third possibility, of course. Instead of seeing animals either as sentient beings or as automata making only fixed reflex responses to stimuli, the neobehaviourist sees them as organisms whose responses vary enormously, through internal as well as external causes. That third choice has been the one adopted by most recent students of migration (see Gauthreaux 1980; Rankin 1985) but it went out along with much else in Baker's "semantic purge" of the subject. He was quite right that something needed to be done about the confused and inconsistent terminology and concepts that had grown up within migration studies as a whole. But the trouble in that field was not a conflict of ideas about behavioural mechanisms but rather the confusion between two distinct levels of migration study, the ecological and the behavioural. On the one hand, migration is an ecological problem of population redistribution by movement (whatever the behaviour), and on the other hand it is an ethological problem of the individual migrant's behaviour, which has been differentiated to varying degrees and in many different forms during evolution (Kennedy 1985*a*). Baker's books are concerned exclusively with the latter.

Baker's view that migrants are sentient creatures carrying a mental image of the right place to be, meant that migration became indistinguishable from movement in general, since it is a truism that animals settle down only in places where some particular stimulus situation obtains. This inference was embodied in his new, stripped-down, undifferentiated definition of migration, viz. "the act of moving from one spatial unit to another" (Baker 1978, 1982). Accordingly, Baker argued that even the well-known, long-range, behaviourally specialized migrations of birds, mammals and fish are mere quantitative extensions of exploratory movements such as occur down to the smallest of scales, and serve to garner information for future, not immediate, use. On this view, these migrants with a "sense of location" do not settle down in a new, distant habitat site on reaching it for the first time. Their first visits are preliminary, purely exploratory visits to that site; the migrants also visit other sites to which they do not return because they do not match their mental image.

> "As it travels, the explorer is continually making comparisons between present and previous locations, rejecting some, ranking others... having visited a succession of sites [it] returns to settle or exploit one of the sites visited earlier in the movement...." (Baker 1982, pp. 52–2)

This is unproved; and hard to credit, if only because at any rate some first-time travellers (e.g. Monarch butterflies going south: ibid., pp. 176–81) simply have not the time to do all the postulated visiting with pauses for side-tracking and back-tracking along the way (ibid., pp. 146–8).

It seems to be true – and hardly surprising – that the classical bird migrants are incapable of homing in on an area from all directions unless they have been there already in a previous year. Nevertheless the travels even of first-time migrants are directionally adaptive. These migrants do not find the right direction by exploratory trial and error. They are well known to head off in the general direction of an appropriate but entirely unfamiliar area by means of what Schmidt-Koenig (1971) called "vector navigation". "Migrant birds employ some type of direction and distance orientation (vector navigation) on their first migratory flight... [but] there is no compelling evidence that first trip migrants are capable of bicoordinate navigation or goal directed homing to the wintering ground. Experience on wintering or breeding areas appears to be necessary to permit homing to those locations in subsequent seasons" (Able 1980, p. 350). Baker (1982, pp. 114, 210–12) called the direction of the vector navigation of juvenile, first-time migrants "a preferred compass direction for exploration". Calling this movement exploratory sustains the fiction that there is nothing special about the behaviour even of long-range migrants: their movements are seen as merely quantitative extensions of local movements. In fact, these migrants do not find the 'right' direction by exploratory trial and error. Their long-range migrations are directional movements using one or other kind of navigation and one or more of a variety of sensory cues, and the key point is that the sensory cues to which the animals are most responsive during migration, unlike those operative in local movements, do not emanate from any suitable settling site (Kennedy 1985*a*). They cannot, therefore, be components of a 'mental image' of one. The migrating animal is far out of range of such local cues for much of the time. The orientation cues employed during this kind of movement are ones that emanate from more general, ubiquitous, environmental features which provide cues that are available en route, such as the sun, or the earth's magnetic field, or the flow of the animal's aerial or aquatic medium. This orientation behaviour differs conspicuously from the variably directed, small-scale, so-called trivial movements made during local foraging and exploratory behaviour by, for instance, the Monarch butterfly in the intervals between its periods of long-range migratory movement (Zalucki & Kitching 1982).

There is another diagnostic feature of a migrant's behaviour that distinguishes it from 'trivial' movement. Responsiveness to the properties of a suitable settling site is continuously present during local movements in and around such a site. But that responsiveness, far from being continuously present throughout a migratory movement as Baker (1982) assumes (see above), is actually more or less inhibited during such movement (Kennedy 1975, 1985*a*). For example Baker (1982, p. 189) writes "we can assume that for most aphids the advantage lies with travelling on the wind at a height just above the highest vegetation... so that they can drop into a potentially suitable habitat as soon as they come across it". A migrant aphid "is wound up to respond to perception of a habitat that matches its innate image of a suitable place to settle, responding to such a match by flying down into that area and beginning to search" (ibid., p. 201). That is erroneous. A migrating black aphid is on the contrary "wound up" to fly and not to respond to a habitat, flying up to great altitudes. It flies repeatedly, precisely because its settling responses are inhibited, even given the stimulus of actual contact with a highly suitable host plant, until it has been flying for some time (Kennedy & Booth 1963). It will respond to the sight of a leaf-green object by flying towards it, but, again, this response is inhibited until the aphid has been flying for some time (Nottingham & Hardie 1989).

It might be argued that Baker is, in effect, following Dennett's (1987) advice and is deliberately describing behaviour as if it were intentional, the better to predict what the animal will do. This pretence is often successful because natural selection will have tended to optimize the animal's behaviour, so that what the animal does will often be what we should consider the best thing for it to do in the circumstances (see **5.1**). However, such anthropomorphic prediction of an animal's behaviour can only be guessing at best and must always be checked directly. If with Baker we believe that animals really do think for themselves like human beings, and act accordingly, then of course checking our predictions about their behaviour does not seem so pressing. Our guesses about what they will do should be a lot better, as good as our guesses about what other human beings will do. Mistakes must inevitably follow such reasoning. Treating animal migrants as sentient beings calculating their movements in advance also means persistently taking the function of their behaviour for its causal mechanism, its ultimate cause for its proximate one (see **3.6**). Altogether, the harm done by anthropomorphism in this case is only too clear.

4.2 Purpose and goal

Quite apart from the recent resurgence of traditional, explicit anthropomorphism discussed in Chapter 2, a new kind of unwitting but implicit and teleological anthropomorphism has appeared. It is the illegitimate child of the post-war development of systems analysis and control theory, especially cybernetics, to which some neobehaviourists turned in order to set up rigorous models of behaviour and to explain it by analogy (Mittelstaedt 1958, 1964, 1978; McFarland 1971, 1989*a*; Toates 1980, 1986; Toates & Halliday 1980; McFarland & Houston 1981).

Toates, who has been a dedicated exponent of this approach to animal behaviour, incidentally exemplifies the way in which neobehaviourists who explicitly disapprove of anthropomorphism in principle sometimes fall into it unwittingly. He is at pains to disavow the anthropomorphic language of earlier times. He objects to its implication of homunculi in nervous systems, of animals with conscious purposes imagining goals and therefore of the reversal of cause and effect as if "time could run backwards" (Toates 1984*a*, *b*). However, he contends that anthropomorphic language has now acquired scientific re-

A purposeful missile? A and B are sensors detecting heat from the exhaust of an enemy aircraft. In the situation shown, A would generate a larger signal than B and would steer the missile in the direction shown by the arrow. The missile's aim would then stabilize where the signals from A and B were equal and the missile therefore pointing straight at the enemy aircraft. (Reproduced with permission from Toates, F. M. 1986. *Motivational systems*. Cambridge University Press.)

spectability. Thanks to the development of engineering control theory, "purpose can be given a physical embodiment" (Toates 1986, p. 11). "We can now believe in goal directed behaviour, without the necessity for time to run backwards" (Toates 1984*a*). Purpose and goal-directedness have their material embodiment, he contends, even in a heat-seeking missile, where any deviations of aim are automatically corrected by negative feedback.

> "It has a signal telling it where it should be (pointing directly at the enemy aircraft), another telling it where it actually is, and the difference steers it to where it ought to be." (Toates 1984*a*, p. 45)
>
> "In one sense, what is a future state, pointing straight at the exhaust of the enemy aircraft, influences its present behaviour... because, in its design, the missile contains a representation of the desired future state. A signal, equality of output of the two heat detectors, embodies the desired future state." (Toates 1986, p. 11)

It is on that reasoning that the equality signal is said to be the physical embodiment of the purpose or goal of the missile.

The description of how the missile works is, however, mistaken. What is embodied in the missile is essentially a system of bilateral heat detectors actuating servomotors that steer the missile in opposite directions. This simple device ensures that pointing at the heat source is the only stable orientation the missile can take up. It is the 'settling point' of the system. It is not a 'set-point', 'reference point', 'reference value' or '*Sollwert*' with which the outputs from the heat detectors are compared, whereupon any discrepancy between them is eliminated by negative feedback, thus keeping the missile pointing at the target. Equal output from the detectors is patently not something that is set independently of the inputs to those detectors. It would have to be independent of them if it was to serve as a standard with which those inputs could be compared. What is signalled by equality of output from the two detectors is their present state, not a future one. The equality of output is not a cause but an effect. It is the effect of *previous* turning carried out in response to previous *in*equality of the outputs. The equal-output 'signal' could only be an antecedent cause of the missile coming to point at the target if time and causation could indeed run backwards (Kennedy 1987*b*).

Thus the engineering analogy does not after all legitimize anthropomorphic, teleological language. Such language remains open to all the old objections which Toates as a neobehaviourist supports. It led him to misapprehend the system. The once-expelled homunculus, conscious purpose, imagined goal and time running backwards (i.e. teleology) have crept in again, quite unintentionally, along with the subjective language now supposedly sanitized by engineering analogy. On the face of it an engineering analogy or mathematical model seems anything but anthropomorphic, but unfortunately it is quite compatible with the anthropomorphic assumptions that we are all prone to make unwittingly (Asquith 1984, p. 168).

Toates's misapprehension of how the missile works is widely shared. R. Dawkins (1976*b*, p. 54), being also a neobehaviourist, gave a guided missile as an example of "the kind of machine or thing that behaves as if it had a conscious purpose", although "nothing remotely resembling consciousness needs to be postulated." Nevertheless he too described the missile as "equipped with some kind of measuring device which measures the discrepancy between the current state of things, and the 'desired' state." Yet the missile contains no fixed standard of comparison representing the "desired" state from which deviations of the current 'state of things' could be measured. Hinde took the same view as Toates and Dawkins: "The behaviour even of a guided missile can be described as goal-directed, in the sense that its course at any moment is corrected according to the discrepancy between its present position and that of the target. It achieves this because some internal model or correlate of the target is compared with its [*the missile's*] present heading or position as indicated by sensors" (Hinde 1982, p. 75). But in fact there are only sensors; there is no internal model or correlate with which the sensors' output could be compared in some way, no *Sollwert*, no goal representation. The only things "measured" and "compared" (figuratively) are the unfixed outputs of the opposing sensors. Expressed more objectively, the sensors by themselves keep the missile on target by means of their opposing outputs to the steering gear (Kennedy 1987*b*). This working principle has long been familiar in the context of orientation under Loeb's (1918) name *Tropism* or Kühn's (1919) revised name *Tropotaxis* (Fraenkel & Gunn 1940; Jander 1970; Schöne 1984; Kennedy 1986).

Hinde (1970, 1982), R. Dawkins (1976*b*), McCleery (1989) and others as well as Toates believe that the mechanism of goal-directed behaviour that they attribute to the missile is widespread among animals.

> "Much animal behaviour can also usefully be described as goal-directed, though precise criteria for when the term can usefully be applied are difficult to specify.... But ultimately 'goal-directed' must imply that the animal has some model or correlate of the goal situation before that situation is achieved, and that behaviour is governed by the discrepancy between current and goal situations." (Hinde 1982, pp. 76–6)
>
> "Cybernetics and automatic guns demystified the concept of goal-directedness in the 40s and 50s." (McCleery 1989)

As Huntingford says, such statements are too sweeping:

> "The concept of a fixed set point on a single dimension, deviation from which controls the start, the strength and

the stopping of behaviour, is too simple for many behavioural contents. Homeostasis may keep the body within certain limits rather than at a precise set point.... Even if behaviour is initiated as a result of a physiological deficit, it may be brought to an end in some other way.... Classical control theory assumes that the performance of the deficit-activated behaviour reduces the tendency to perform that behaviour; however, behaviour is often self-reinforcing." (Huntingford 1984, pp. 82–3, 85)

McFarland (1981), who has contributed much to the application of control theory to behaviour, remarked that the danger of the discrepancy-correcting approach was "that the behaviour may come to be described in unacceptable teleological terms". More recently McFarland (1989*a*) has drawn renewed attention to the classical alternative of a Tropotaxis-like mechanism that could explain cases of apparently goal-directed behaviour without invoking some correlate, set-point, *Sollwert*, model, image or other representation of the goal. The goal-directed mechanism has often been assumed without the 'tropotactic' alternative having been ruled out; moreover we have seen that the guided missile and various animals conform to the latter. McFarland (1989*a*) calls this alternative mechanism "goal-seeking" behaviour to distinguish it from strictly "goal-directed" behaviour (it is a pity he chose to use the anthropomorphic term "seeking": see p. 163). To illustrate the strictly goal-directed mechanism he took a physiological model, the thermostatic theory of body temperature regulation in animals, and usefully spelled out the model's properties in full in order to contrast them with mere "goal-seeking":

"(a) There is an internally represented reference temperature (often called the set point) with respect to which the body temperature is judged to be too low or too high,

> (b) there is a mechanism (called a comparator) for comparing the set temperature with the body temperature, (c) the output of the comparator actuates the heating and cooling mechanisms." (McFarland 1989*a*, p. 109)

The alternative, so-called goal-seeking, theory, is that

> "the processes controlling heating and cooling balance each other over a wide range of conditions without any representation of a set-point. This alternative view has led some to argue that the set-point concept has little more than descriptive value ... the goal achieved by an animal is often an emergent property of the system, and ... it is a mistake to assume that the goal must necessarily have a representation in the animal .... This kind of argument can be applied at many levels, such as the level of peripheral control, of orientation, of physiological regulation or of motivation." (Ibid.)

Thus, in McFarland's terminology the guided missile could be called a 'goal-seeking' system but not a 'goal-directed' one. "A goal-seeking system is one which is designed to seek the goal without the goal being explicitly represented within the system. Many physical systems are goal-seeking in this sense. For example a marble rolling round in a bowl will always come to rest at the same place. It ... appears to be goal-seeking because the forces acting on it are so arranged that the marble is 'pulled' towards the goal. In some goal-seeking systems that are designed to maintain a particular level, or direction, a dynamic equilibrium is maintained by self-balancing forces .... A goal-directed system ... is directed by reference to an internal representation of the goal-to-be-achieved .... In this category are included the 'set-point' of simple servomechanisms, the 'Sollwert', or 'search image' of classical ethology, and any form of explicit mental representation as postulated by cognitive ethology and some branches of psychology" (McFarland 1989*a*, pp. 106–7).

In the event, McFarland (1989*c*) finds "there is no hard evidence for explicit representations that control behaviour in the manner of a 'set-point', or 'sollwert'" (cp. Wirtshafter & Davis 1977; Cecchini *et al.* 1981). They are on a par with 'search images' (see **3.3**). All this is not to say that 'goal-directed' systems in McFarland's objective sense do not exist. The point here is only that many neobehaviourists have assumed that animals are using such a system even when the hypothesis that the animals are using a simpler, 'goal-seeking' system has not been ruled out and this latter looks, in fact, more likely.

Schöne (1984) differs from the other writers on orientation mechanisms cited above in his strong commitment to the 'goal-directed', '*Sollwert*' or telotactic model as against the 'goal-seeking' or tropotactic one. He presents valuable criticisms of some of the criteria conventionally used to distinguish the tropotactic model from the telotactic one. But the refrigerator which he takes as his physical model is like the guided missile in that it embodies no material entity representing the *Sollwert* (or "reference value" as he calls it here; ibid., p. 50). The system is a 'goal-seeking' one. Schöne (ibid., p. 21) makes the important point that the external stimulus that initiates an oriented movement always has the additional function of setting the direction that the movement will take with respect to the orienting stimulus (the initiating and orienting cues may be different, as when an odour initiates a movement that is oriented on the wind). Natural selection will of course have settled what that direction will be as an outcome or ultimate cause of the behaviour, but Schöne gives it a proximately causal role as well. He defines it as the reference value of the orientation, thus simply assuming that the movement must be 'goal-directed', although that is not something that can be assumed.

4.3 Efference copy

The suppression of the compensatory optomotor reaction when an animal moves is another case where it has been supposed that an internal representation or image of the 'expected' outcome of a movement serves as a standard for comparison with current sensory input. When the whole visual surroundings of an animal move, the animal is stimulated to turn in the same direction, thus tending to compensate for the movement of images over the retina induced by the surround movement (negative feedback). That is the classic optomotor (or optokinetic) reaction. If the surround remains still and the animal itself makes a turn (or turns its eyes) this too must induce image movement over the retina. Yet this does not stimulate a compensatory turn: the animal is able to proceed with its turning unhindered. At one time this was taken to mean that the optomotor reaction was simply inhibited during self-movement, but Mittelstaedt (1949; von Holst & Mittelstaedt 1950; English accounts in von Holst 1954, 1973; Mittelstaedt 1964) disproved that assumption in experiments on a hoverfly walking inside a vertically striped drum. If the fly's head was rotated on its slender neck through 180° and fixed there, the optomotor reaction to rotation of the drum was reversed but unimpaired. The direction of image movement over the eyes being now anatomically reversed, the drum rotation now naturally stimulated the fly to turn against the direction of the surrounding stripe movement instead of with it. The significant point was that before the eyes were transposed the fly had been able to make 'voluntary' turns perfectly freely in the stationary drum with no hint of opposition from compensatory responses. This was what gave rise to the presumption that optomotor responses were simply inhibited, shut down altogether, during such turns. It became evident that this was not so when the eyes were transposed. For now, as soon as the fly initiated a 'voluntary' turn (in a stationary drum) the turn was promptly magnified into rapid, tight spinning one way and the other until the fly was exhausted. This meant that the optomotor reaction was occurring, after all, because, the image movement being now reversed, optomotor responses would reinforce the image movement instead of counteracting it, generating positive instead of negative feedback.

To account for this behaviour Mittlestaedt and von Holst postulated that when the central nervous system initiates a turn by issuing an appropriate set of commands to the leg muscles, an image or copy of this 'efference' is retained in the central nervous system. As the commanded turn proceeds the sensory feedback or 're-afference' from it compares to the efference copy "as the negative of a photograph compares to its print; so that, when superimposed, the image disappears" (von Holst 1954). In other words, if the afference matches the efference copy, then the afference must consist solely of re-afference due to the fly's turning itself, and the retained efference copy will nullify it. The fly's self-initiated turning can then proceed unhindered. If, however, there is any discrepancy or mismatch between the afference and the efference copy then not all of the afference can be re-afference from the fly's turning movement. Some of it must be 'ex-afference' from real movement of the surround, and this is liable to interrupt the self-turning by evoking a compensatory optomotor response. With the eyes transposed any turning, whether to the left or the right, itself produces a mismatch which optomotor responses can only worsen.

This experiment and interpretation have often been held up as a model system (e.g. Tinbergen 1969; Hinde 1970, pp. 97–9; Evarts 1971; Teuber 1974; Stillar 1985) because "any movement made by an animal produces consequences which must be distinguished from independent changes in the external environment" (McFarland 1986). McFarland's account will serve to illustrate the generally accepted version of the von Holst/Mittelstaedt theory:

> "Motor commands ... not only cause patterns of muscular movement, but also set up an *output copy* [efference copy], which corresponds to the expected input from the sensory processes that monitor the limb movements of the animal. The brain then makes a comparison between the output copy, i.e. the pattern of signals sent out by the brain, and the incoming or *afferent* information, i.e. the pattern of signals transmitted to the brain by these sensory processes.... If the brain finds no differences, then it concludes that all the incoming information was afferent. This means that all the instructions given by the brain ... were carried out. If, however, the comparison shows a discrepancy between the output copy and the incoming information, then ... exafferent information was also received by the sensory processes involved." (McFarland 1981*a*, p. 469)

Von Holst (1954) brought forward long-known human subjective experiences with eye movements as evidence that the postulated efference copy really existed, and referred to Sperry's (1950) analogous postulate of a "corollary discharge". These experiences have been summarized by Evarts. "We assume that each voluntary movement, or change of posture, involves not only a downward discharge to the peripheral effectors, but a simultaneous central discharge from motor to sensory systems preparing the latter for those changes that will occur as a result of the intended movement. A voluntary eye movement which transports contours across the retina would leave the spatial order of perception undisturbed, because the impulses to the eye muscles are accompanied by appropriate corollary discharges which preset the visual system for all anticipated shifts in the spatial order of visual inputs. By contrast, when we push against our eyeball, moving it passively, the visual scene jumps [in the same direction], and the same apparent visual shift of scene is perceived whenever we intend to move our eyes but there is an inability to move [them].... These illusory motions...provide the best direct evidence for the continual operation of the compensatory mechanisms which counteract the normally inevitable shifts of input that result from voluntary movement" (Evarts 1971, p. 107).

Since then some direct and clear neurophysiological evidence of the postulated discharge causing 'saccadic suppression' in the locust has been obtained by Zaretsky & Fraser-Rowell (1979). Less direct evidence concerns certain neurones in the monkey's superior colliculus that fire regularly when the eyes are stationary and vertical stripes are moved in front of them. When the stripes are stationary and the eyes themselves are moved these cells receive a signal, despatched not from the retina but probably from the eye muscles, and fall silent (Robinson & Wurtz 1976). This zero discharge period is presumably a consequence of the inhibitory intervention of the efference copy signal, an inhibitory pattern precise enough to delete any trace of optomotor responses. In addition, discharges synchronous with spontaneous eye movements and having no discoverable function other than that of efference copies have ben recorded from the tectal commissure of goldfish (even after removal of the tectum) (Johnstone & Mark 1979; see also Paul 1989).

It will be apparent that the efference copy is equivalent to the *Sollwert* of the previous examples (see **3.2**, **3.3**) of goal-directedness supposedly based on negative feedback from any discrepancy between the *Sollwert* and the current sensory input here called the afference. There is evidence that corollary discharges and re-afferences providing for negative feedback really exist. But the idea of the behaviour being guided by comparison of the afference with a copy of the motor command is as questionable here as in the previous cases. Mittelstaedt (1971) himself eventually decided it was most unlikely; novel and more thorough experiments with Colorado beetles led Lonnendonker & Scharstein (1991) to conclude "a Colorado beetle has no internal representation of its spontaneous long-term turning tendency". My critique that follows may be related to MacKay's critique of the Mittelstaedt/von Holst theory, which Mittelstaedt accepted (MacKay & Mittelstaedt 1974).

Motor commands contain both excitatory and inhibitory signals. Effective action usually requires both excitation of one set of muscle contractions and at the same time inhibition of antagonistic ones. Mittelstaedt's experiment showed clearly that optomotor responses were not *completely* inhibited during self-movement, contrary to the *Reflextheorie* of the radical behaviourists, which he and von Holst were evidently keen to refute. However, having made that point, they made no further mention of nervous inhibition and pursued their explanation in non-physiological, metaphorical terms more appropriate to human ratiocination: comparison, expectation and so on. Subsequent authors have continued to recount the story in that way, effectively blocking any further physiological analysis. But the experiment did not rule out *partial* inhibition of the optomotor responses. Physiologically, we should expect there to be selective inhibition of any responses that were antagonistic to the commanded movement. The inhibitory component of the command, delivered alongside the excitatory component, would ensure that the commanded self-movement could then proceed without hindrance from optomotor responses. Moreover this would be a continuously running process as the movement proceeded with fluctuations of the pattern of excitations and inhibitions required for it. Unlike comparison, expectation, etc., it would also invite neurophysiological analysis.

Thus the efference copy cannot be regarded as an 'image' filed away in the CNS, a special entity (like a 'search image') additional to the ordinary integrative activity of the CNS. Nor does the efference copy need a comparator to be able to guide behaviour. Indeed, it is not always necessary for the inhibition of motor responses antagonistic to a commanded self-movement to be deferred until some re-afference comes in and can be compared with the efference copy. Evarts (1971) mentions reports dating from the beginning of the century that "a number of different sorts of movement are not controlled by sensory inputs during their execution". They are controlled not by negative feedback from the muscles but by negative "feedforward" within the nervous system; or, in other words, by a corollary discharge that is unaided by re-afference. In the monkey, for example, "cerebellar neurons ... discharge prior to learned movements and well in advance of any response feedback"; and in people, the tracking error correction time was found to be less than the proprioceptive reaction time (Evarts 1971; see also Schöne 1984, p. 52).

To sum up, what seems to have happened in this case is that the physiological baby has been emptied out along with the radical behaviourist bath water. Rejecting the radical behaviourists' crude conception of inhibition as an all-or-none process in reflex integration, von Holst & Mittelstaedt (1950) and later writers nevertheless acted upon it, in effect, by ruling out any inhibition at all upon finding that the optomotor responses were not *totally* inhibited during self-movement. They overlooked the likelihood of *some* role for nervous inhibition in the partial disappearance of such responses and resorted instead to the non-physiological, subjective concepts that we employ in conscious analytical thought (cp. p. 113).

This anthropomorphism made a readily acceptable story at the cost of a mistake. Also, and as before (p. 70), engineering control theory will have favoured the idea of an efference copy that is compared with the re-afference.

4.4 Motivation

Lorenz extended Descartes's dualism from people to animals in the original Grand Theory of instinct (**3.1**). He described the "energy" or motivation for any action, which came from within, on the one hand, and the reflex machinery that executed the action, on the other, as "two absolutely heterogeneous causal factors" (Lorenz 1950, p. 251). He saw the "endogenous activity as a distinct physiological process ... an independent, particulate function of the central nervous system ... equally as important as the reflex" (ibid., p. 249). The conception of motivation that is accepted among ethologists now is on the face of it radically different, a unitary concept. In their up-to-date dictionary, Immelmann & Beer (1989) defined motivation simply as "a general term to cover the subject of proximate control of behaviour". Halliday (1983, p. 105) expressed this more fully: "At best, intervening variables are labels for unknown physiological process .... The ultimate aim of much research into motivation is to identify and understand how such processes work, so that concepts such as hunger, thirst and drive become unnecessary."

Yet the following two equally representative statements show that Lorenz's dichotomy has not been sunk without a trace:

> "There is no difference of kind between simple reflexes and more complex reactions ... but ... applying the term 'reflex' in such a broad manner makes it synonymous with 'behavior', which does not help us at all. It is

customary, therefore, to restrict the term 'reflex' to relatively simple and automatic responses to stimuli and to designate more complicated behaviour patterns by other terms." (Keeton 1967, p. 45)

"In reflexes, there is little more than a simple neural pathway and a clear-cut, often momentary, response. Most other behaviour involves several complicating additions: the animal's motivational state, hormone levels, physiological rhythms, cognitive processes ... and so on." (Barnard 1983, p. 27)

At first sight those statements look like plain common sense. But they imply, albeit unintentionally, that there is in fact a difference of kind between reflex behaviour and 'motivated' or 'cognitive' behaviour, after all (Kennedy 1987*a*). We shall see in **5.2** and **6.1** that in practice the psychological provenance of those two terms can lead to motivation and cognition being regarded as things exclusive to the highest levels of the causal hierarchy, not as physiological processes. That there is a persistent tendency to Lorenzian dualism can be inferred also from the trouble Gallistel took to argue in the opposite sense: "the problem of motor coordination becomes the problem of motivation as one ascends the action hierarchy" (Gallistel 1980, p. 287). "The principles governing waxing and waning of potentiation at motivational levels in the hierarchy are much the same as those encountered at lower levels" (ibid., p. 332).

This matches the conclusion arrived at by the insect physiologist Dethier after many years of experimental work on the behaviour and physiology of feeding in a fly. Interestingly enough Dethier started out convinced that there were indeed "motivational aspects of instinctive behaviour" and that it was "quite meaningful to talk about 'drives', 'goal-directedness' and 'satiation'" (Dethier & Stellar 1961, p. 80). The long-term research aim he set himself was "to ascertain whether or not it [*motivation*] occurred in an organism whose evolutionary appearance predated that of man, to discover whether or not its expression lay within the capabilities of a relatively simple nervous system, to enquire into the reality of motivated behaviour as a separate class" (Dethier 1966, p. 128; see Kennedy (1987*a*) for a resumé of this saga). In the event Dethier could find no evidence to satisfy a physiologist that motivation was a separate class; and that not only in insects but also, after a thorough survey of the extensive literature on motivation in general, in any animal. He was finally driven to the conclusion that "the concept of motivation ... has not only outlived its usefulness ... but has become an impediment to our understanding of the behavior it purports to explain" (Dethier 1982).

It has been argued that, for the time being, we need 'motivation' as a blanket term to cover the collective causes of any behaviour because "we do not wish to wait around for the several hundreds of years that physiologists are going to take to remove our ignorance" of "things like interactions between nerve cells and the release of hormones" (M. S. Dawkins 1986, p. 94). But there is no point in our waiting for all that low-level physiological information to be collected since such a mass of detail "will only confuse and make it more difficult to see how the whole works", as M. S. Dawkins (see p. 126) has well said.

Even if we do want a blanket term, there does not seem to be any good reason for going outside our discipline for it. So the question arises, why is it that the causation of whole animal behaviour has come to be called by an imported, non-physiological name – motivation? The only apparent reason is that it is of course only at the level of the whole behaving animal that we feel empathy towards animals and are addicted to imputing human mental processes to them. So it seems to us perfectly natural to borrow a term such as 'motivational' from human psychology for use at the whole-animal level. At that level the individual animal seems to all intents and purposes a person to us especially, of course, if it is a pet. In our everyday thinking the overall behaviour of a human person is not physiology but the work of a mind. Only a person of 'sound mind' is deemed to be legally responsible for his or her actions. Unlike Descartes, we still tend to think of the behaviour of an animal in a dualist way as something separate from its physiology – as if the behaviour of the animal, too, were the domain of a mind (see also **5.4**).

One effect of that is to open the door to cognitive interpretations of animal behaviour and the mistakes they can generate (**5.2**). The more general consequence of using the subjective term 'motivation' to designate the causal mechanisms of behaviour is that it unwittingly brings back to life the discredited notions of a 'linear causal chain' or 'unitary drive' (see **3.1**). It does this by designating the outcome of the causal process behind a given behaviour as its antecedent cause, mistaking its effect for its cause (see **3.6**).

CHAPTER 5

Chapter 3 included two generally recognized errors (see **3.4** and **3.5**) that did not spring from teleology but from another form of anthropomorphism, namely ascribing our cerebral abilities to animals. This chapter presents three other cases where it is unjustifiably assumed that the animals share these abilities, and one case of assuming that they share our feelings.

5.1 Intentionality

We have intentions, and we tend to assume that animals have them too, which we cannot know. To the layman 'intentionality' sounds like nothing more than a learned synonym for purposiveness (**4.2**), but it has come in for a good deal of discussion recently especially among philosophers and psychologists. For instance, Asquith queried the intentions imputed to higher primates by workers who describe these animals as giving each other "warning" signals: "'Warning' conspecifics about a possible predator may be recognized (defined) by the function that the behaviour fulfils regardless of how it was manifested (for instance, by shaking branches, calling, running or freezing). However... the animal's intention to perform the function... is not scientifically testable" (Asquith 1984, pp. 142–3). Intentionality is indeed scientifically untestable in animals, but among people warnings are, by definition, intentional. Hence describing an animal's actions (shaking branches, etc.) as a warning to its fellows effectively takes it for granted that those actions were intentional on the part of the animal. Intentions are proximate causes of behaviour, so again we have an ultimate, functional cause masquerading as a proximate one in the animal (cp **3.6**).

Before Darwin, as we all know, it was not realized that the striking adaptedness of behaviour was the product of natural selection; and the great swing of opinion that followed was accurately summarized by Sherrington, writing at the turn of the century:

> "Older writings on reflex action concerned themselves boldly with the purposes of the reflexes they described. The language in which they are couched shows that for them the interest of the phenomena centred on their being manifestations of an informing spirit resident in the organism.... Progress of knowledge tended more and more to unseat this anthropomorphic image of the observer himself which he projected into the object of his observations." (Sherrington 1947, pp. 237–8.) "That a reflex action should exhibit purpose is no longer considered evidence that a psychical process attaches to it, let alone that it represents any dictate of 'choice' or 'will'. In the light of the Darwinian theory every reflex must be purposive." (Ibid., p. 236)

Although Sherrington thus made it quite clear close on a century ago (Sherrington 1947, reprinted from 1906) that "purposive" had now become merely a metaphor for adaptive, the term purposive was not dropped but remained in use with and without inverted commas. Inevitably it was not entirely divested of its everyday mentalist overtones. Because of them, observers easily misinterpret evidence of adaptedness as evidence of purposiveness, E. C. Tolman being a classic case (Boakes 1984, p. 236). Since a purpose, like an intention, is a proximate cause of our behaviour as we see it (*pace* McFarland, p. 30) we assume that it is a proximate cause of animal behaviour as well. That is what Dunbar assumed when he said:

"to be able to study these kinds of social systems the observer has to second-guess what his animals are up to ... we need to determine not only what an animal does, but, more importantly, what it is *trying* to do" (Dunbar 1984*a*, p. 231). To assume that an animal is trying to do something is unwarranted anthropomorphism, although empathy makes the idea terribly hard to resist, especially with primates, and Dunbar for one did not resist it. Convinced that what he was doing was "thinking himself into the animal's 'state of mind'" (ibid.; see also pp. 5 and 26), he interpreted its behavioural adaptations as its intentions, confusing ultimate and proximate causes of behaviour. His anthropomorphism was unwarranted because it is natural selection and not the animal that ensures that what it does mostly 'makes sense', as we are wont to say. We intend our own behaviour to make sense, meaning to serve us well, and if we can make sense of an animal's behaviour it looks intentional to us by analogy. All that this really means is that the bit of behaviour in question fits in well with the rest of the animal's behaviour. All the items in an animal's behavioral repertoire are co-adapted and integrated into a single adaptive package – a "great unitary harmony" Sherrington (1947, p. 238) called it in his poetical style. Or in McFarland's (1989*b*, p. 46) more prosaic words, "We can expect natural selection to shape the decision-making mechanisms of animals in such a way that the resultant behaviour sequences tend to be optimally adapted to the current situation." Or in Gallup's words (1982, p. 247), "organisms have evolved in many instances to act as if they had minds". That is not to say that animals have minds, but it is what makes it profitable in practice to assume for argument's sake that they have minds – just so long as we do not take this literally, as Dunbar did, which would be genuine anthropomorphism.

Assuming, just for argument's sake, that an animal is behaving purposefully is of course the rationale of modern studies of the functions of behaviour under the name behavioural ecology. This new discipline has effected a major breakthrough by introducing the rigorous quantitative study of behavioural adaptation. Even short of that, observers working on the assumption that animals have minds can often make reasonable guesses as to what an animal will do in given circumstances, by considering what they themselves might choose to do if they had the same capacities and constraints as the animal. The fact that we are able to do that is sometimes called (rather cryptically) the 'heuristic value' of anthropomorphic teleology; and it is extremely useful when seeking the function of behaviour. "Anthropomorphizing *works*" declared Cheney & Seyfarth (1990, p. 303) from their long field experience with monkeys: "attributing motives and strategies is often the best way for an observer to predict what an individual is likely to do next". Or Dunbar (1984*a*, for example, continuing his p. 231), says that "unless this can be done, it will often not be possible to appreciate the options that are open to the animal". 'Options' for Dunbar seems to mean deliberate choices by the animal, but it can simply mean alternative ways of reacting to the same stimulus when circumstances differ, alternatives with which natural selection has equipped the animal. It is not necessary to be an outright anthropomorphist like Dunbar to profit by our own thoughts about what the animal could and should do. For a neobehaviourist it is enough to pretend to be an anthropomorphist, merely imagining what adaptive reactions the animal is likely to have been equipped with by natural selection. Thus de Waal (1989) also finds 'heuristic value' in an empathic approach to monkeys and apes even though, as we saw (p. 27) he is not altogether convinced that they are capable of thinking out for themselves the best course of action to take.

In fact all students of whole animal behaviour regularly practise such 'mock anthropomorphism', by imagining what an animal is 'trying' to do, or guessing what it will do on the basis of what we would think it best for it to do. In this way we can guess the function of its behaviour. That is to say we habitually anthropomorphize about animal behaviour, using our own mental processes as models to 'explain' the behaviour in terms of intentions; and again, this is very useful. "Denying oneself the use of empathy, intentional language and the associated concepts has real disadvantages in terms of achieving understanding about behaviour" (Bateson 1990). By "understanding" Bateson presumably means understanding the function of some behaviour, not its mechanism. Like any analogizing, anthropomorphizing helps us to "get our minds round" (ibid.) complex processes, in this case those of intact behaviour (e.g. primate social behaviour).

But those formulations are too vague. They could be taken to mean be that anthropomorphizing helps us to understand not only the ultimate causes or functions of animal behaviour, but also its proximate causes including 'intentions'. This was the view of Wilder (1990), for example, and it would of course be genuine anthropomorphism which Bateson presumably did not intend. Krebs & Davies (1987, quoted on p. 52) avoided any ambiguity here, and so did Asquith (1984, p. 164; my italics), who said "ordinary language terms can be heuristic aids to theory construction about the *function* of behaviour" while at the same time warning that our ordinary language descriptions of animal behaviour employ many terms that "carry meanings associated with human action (that is, purposeful behaviour)" (ibid., p. 138). We habitually resort to anthropomorphic metaphors also when we describe the complex behaviour of inanimate systems such as the weather or computers, but there is little danger that we shall take the analogies literally. This use of mock anthropomorphism runs no risk of being mistaken for genuine anthropomorphism. Mock anthropomorphizing about whole-animal behaviour itself is crucially different. It does run that risk, suggesting that we have really explained the behaviour in a proximately causal sense when of course we have done nothing of the sort. It is veital to keep in mind, also, that our predictions of an animal's behaviour based on anthropomorphism are no more than hypotheses that need to be tested. And if the tests bear out our guesses this does not mean that the animal itself acted cognitively, after taking thought. This point was neatly put by Bateson (1990): "attributing the power of making choices to an animal, so that we can do more imaginative science, does not mean that, when our efforts are crowned with success, we have proved that the animal has chosen.... When we find it helpful to suppose that animals have preferences, the way we think about them is not evidence that they think."

"Imaginative" seems again an unfortunately vague term but, that aside, Bateson was there illustrating the distinction between 'mock' anthropomorphism, which refers to ultimate causation, and genuine anthropomorphism which refers illegitimately to proximate causation: a distinction which is as difficult to maintain as it is crucial. Neoanthropomorphism means being genuinely anthropomorphist implicitly, but unintentionally. Thus neobehaviourists are not immune from neoanthropomorphism, because it is unconscious. It is often hard to tell whether an author's anthropomorphic language is of the mock or the genuine variety, or unthinkingly ambiguous. Some of such ambiguous cases appear in **5.3**. The expression "behavioral awareness" used by Savage-Rumbaugh and quoted on p. 113 is particularly bizarre, for of course behaviour as such cannot be aware. The awareness here is no more than a hypothetical mental state inferred from the behaviour by human analogy: straight anthropomorphism. This expression recalls the chimerical grafting together of human-psychological and physiological terms that Tinbergen (1951) once coined in "motivating impulses". He abandoned it later and had in fact criticized confusion of psychological and physiological thinking in the very same article (see p. 51, above).

Nevertheless this mixing of immiscibles continues to be widespread among those neobehaviourists who have some regard for animal cognitivism. An important recent exemplar is Cheney & Seyfarth's thoughtful account of their richly productive field work on the social behaviour of vervet monkeys. They conclude that the monkeys are not actually aware of their complex social relationships or of their postulated mental representations of them. "Though a monkey may make use of abstract concepts and have motives, beliefs, and desires, her mental states are not accessible to her: she does not know what she knows" (Cheney & Seyfarth 1990, p. 312). Yet the authors routinely employ subjective terms, writing about what the monkeys *know*, what they do with this *knowledge*, what goes on inside their *minds*, their various *mental states*, etc. – without inverted commas. Now we have of course no difficulty in recognizing that there are different levels of knowledge in ourselves (cp. Weiskrantz's idea (p. 19) of our consciousness as a monitoring system with "varying levels of abstraction in thought"), but we can impute any knowledge to animals only by intuitive anthropomorphism (p. 27).

Cheney & Seyfarth (1990) accept that "descriptions of social behavior in anthropomorphic terms do not constitute an explanation" (ibid., p. 303) and even mention an alternative hypothesis: "it is unclear precisely how the monkey's representations might differ from associations formed through classical conditioning" (ibid., p. 175). For that reason, they explain, a second goal of their book is "to dissect the knowledge and motives that make monkeys do what they do" (ibid., p. 303). This plainly does not refer to the ultimate causes of the monkeys' behaviour, causes to which subjective concepts such as knowledge and motives are commonly applied as mock anthropomorphism. It refers unequivocally to the proximate causes: e.g. vervet calls "are caused by the mental states of those who use them" (ibid., p. 312). This is not mock anthropomorphism but unwitting or even explicit anthropomorphism.

A non-primatological exemplar of mixing immiscibles is McFarland's (1985) definition of the motivational state of an animal: "The behaviour we see is determined by the brain in accordance with [the animal's physiological] state and in combination with the animal's perception of environmental stimuli. The combined physiological and perceptual state, as represented in the brain, is called the 'motivational state' of the animal" (McFarland 1985, p. 279). Perception is a psychological term and its non-physiological nature is underlined by its being presented in the passage above as something separate from the animal's physiology. Yet the animal's motivational state is said to be made up by somehow combining this ostensibly non-physiological process with the animal's physiological state.

It is particularly hard to discern the position of the ethological philosopher Dennett, who seems perhaps to be embracing genuine anthropomorphism like Dunbar, albeit more obscurely and at prodigious length. For example: "Even if we feel comfortable attributing a belief to a frog ... there are apparently no principles available for rendering the content of the attributed belief precise.... And yet this anthropomorphizing way of organizing and simplifying our expectations about the frog's next move is compelling and useful" (Dennett 1987, p. 108). Or again: "imagistic characterizations are perspicuous because they are richly predictive of a surprisingly wide variety of behavioural effects. Talking of mental images may be a *façon de parler*, but it is no mere *façon de parler* because taking the talk (quite) literally keeps on leading to confirmed predictions" (Dennett 1978).

Thus Dennett admits there that ideas based on anthropomorphizing, such as the idea of frogs having beliefs and mental images, cannot be proved, but marvels at how successfully predictive it is to postulate them. It has been Dennett's principal concern to bring out the heuristic value of adopting what he calls the "intentional stance" when studying animal behaviour. By an intentional stance he means assuming for argument's sake that the animals are acting intentionally. It is equivalent in other words to mock anthropomorphism: Dennett does not embrace genuine anthropomorphism in so many words like Dunbar. Nevertheless, in McFarland's (1989*c*) opinion Dennett's reasoning implies genuine anthropomorphism. Consistent with McFarland's claim is the way Dennett repeatedly writes (as in the above passages) as if the great predictive power of adopting the intentional stance were so surprising that it should make us all think twice before dismissing "belief talk and desire talk" about animals as a mere metaphor or *façon de parler*. But this predictive power should come as no surprise to us as we saw above. It was acknowledged long ago by H. S. Jennings (1906): "We do not usually attribute consciousness to a stone... it would lead us much astray in dealing with such an object. On the other hand, we usually do attribute consciousness to the dog... it enables us to appreciate, foresee, and control its actions much more readily than we could otherwise do so." Nowadays, recognizing that natural selection is an optimizing force, we do not hesitate to hazard a guess as to what an animal will do, based simply upon what seems to us would be best for it to do; and it will often be a good guess. But the beliefs and desires of animals remain purely metaphorical.

This ability of ours to make good predictions on the basis of mock anthropomorphism is the point that Lockwood (1985/86, p. 188), for example, entirely missed in elaborating his thoughtful defence of anthropomorphism against behaviourism. "For the last 100 years", he wrote, "the implication has always been that anthropomorphism is bad science.... A primary purpose of scientific method is to enable us to make valid predictions about the world.... If anthropomorphism generates bad science...then that should mean that the hypotheses we generate based on assuming things like emotion, intelligence, feeling, intention and so on, will be very inaccurate." Lockwood is by no means alone in not having caught up with the fact that some anthropomorphism need not generate bad science, and that is *mock* anthropomorphism, because natural selection has produced animals that act *as if* they had minds like us. That is what enables us to make valid predictions about animals' behaviour without making unprovable assumptions that they have feelings and intentions. Genuine anthropomorphism does make those assumptions and is therefore something quite different. It assumes that animals have mental events as we do and sees these events as proximate causes of their behaviour as they are of ours. For that reason it does often generate bad science.

It is common even today for students of the field behaviour of animals to miss the vital distinction between genuine and mock behaviourism. In this confused situation we find the intuitive kind of anthropomorphism known as cognitivism (see **5.2**) taking over. De Waal's and Dunbar's statements about the intuitions of primatologists (pp. 27–8), and what has been said just above about predictability, are strikingly confirmed in a report by A. Wannenburgh about Bushmen and their prey animals:

> "Bushmen acquire an extensive knowledge of animal behaviour through constant observation, careful attention to detail and continual discussion among themselves of what they have seen. Their understanding enables them to identify completely with the animal they are hunting, so that they can answer such questions as: 'What should I do now if I were this animal?' And their

> replies are amazingly accurate. Such anthropomorphic projection is rejected by scientists, who say it is not possible to equate human and animal consciousness." (Wannenburgh cited by Fox 1984, p. 132)

In the first part of that passage, Wannenburgh gave the reason why many scientists, so far from flatly rejecting anthropomorphism as he claimed in the last sentence, make regular use of it, whether or not they realize that it is only the mock form of it that they need for their purpose. He, like Lockwood above (and Fisher 1991), does not recognize that it is only anthropomorphism in its genuine form that scientists need to reject, because it assumes animals are conscious.

Turning back to Dennett, he seemed to be arguing in another place in his book that neither animals nor people act intentionally (not that both act intentionally which is the anthropomorphic position) when he wrote, "there is nothing more to *our* having beliefs and desires than our being voluminously predictable (like the frog, but more so) from the intentional stance" (Dennett 1987, p. 108). "We human beings are only the most prodigious intentional systems on the planet, and the huge psychological differences between us and the frogs are ill described by the proposed contrast between literal and metaphorical belief attribution" (ibid., p. 112). It is not obvious what is wrong with that description of the differences between us and frogs; the "nothing more" in the first sentence seems to place Dennett, in effect, in the position of a radical behaviourist dismissing introspection. If we do not dismiss it, then we have to report that beliefs, desires, intentions, images, etc., are real features of our conscious lives. They are literal descriptions of experience and not abstract metaphors, whatever their physiological nature may be.

Bonner's (1980) elegant and important book on the evolution of culture in animals presents us with a difficulty of a different kind. He recognized the danger of anthropomorphic descriptions of animal behaviour in virtually the same terms as this book: "It can be argued that no matter how excellent and pure our stated intentions might be, the words will unconsciously tend to make us interpret animal behavior in human terms" (Bonner 1980, p. 12). And yet I cannot comprehend his view that the behaviour of animals carries, if anything, less danger of anthropomorphic misinterpretation than the behaviour of inanimate systems. He said "one might suppose it is easier to separate Newtonian mechanics from our psyche than courtship and altruism in the behaviour of birds, but in fact they are both seen through our minds. If anything, in the behavior of birds it is possible to see the pitfalls simply because they are more obvious" (ibid.). Admittedly we all 'anthropomorphize' without hesitation about inanimate systems such as hurricanes and ships and computers. But this does not matter. It does no harm because, *pace* Bonner, people in our culture are not likely to take it literally and imagine a purposeful being animating the inanimate system. Here, "any anthropomorphic terms such as 'Maxwell's Demon' are unmistakably metaphors only" (Kennedy 1986, p. 23). By contrast, anthropomorphizing about animal behaviour is easily taken literally.

5.2 Cognition

Over the last two decades, alongside and often combined with the massive growth of behavioural ecology with its routine use of anthropomorphic language, there has been a substantially increased use, among animal psychologists and philosophers particularly, of what are called cognitive interpretations of animal behaviour (Hulse *et al.* 1978; Mellgren 1983; Roitblat *et al.* 1984; Dickinson 1985; Terrace 1985; Roitblat 1987; Bekoff & Jamieson 1990*a*, *b*, 1991).

Bekoff & Jamieson gave an indication of how cognitive ethology was born from an engineering analogy (Cp. Toates on goal directedness on pp. 67–70):

> "Just as it is often appropriate to explain the behavior of a computer in terms of its program, so Putnam suggested it is reasonable to explain human behavior by reference to mental states. By assimilating mentalist language to the program states of a computer... we could be both mentalists and materialists... By the 1970s the "cognitive revolution" was upon us.... It was no longer out of the question for scientists to explain human behavior in mentalistic terms... then it seemed natural to explain animal behavior in mentalistic terms as well" (Bekoff & Jamieson 1991, pp. 4–5).

Definitions of cognitive psychology and ethology vary widely but a common feature is that "in contrast to behaviourism, cognitivism focuses on the internal representation of knowledge, and insists that theoretical concepts such as attention, expectancies, images, intentions, goals, plans, and templates are essential for the understanding of behavioural patterning" (Colgan 1989, p. 63). Cheney & Seyfarth's position fits that definition (pp. 91–2) although they give H. Markl's more objective-sounding one: "the ability to relate different unconnected pieces of information in new ways and to apply the results in an adaptive manner" (Cheney & Seyfarth 1990, p. 9). However, *unconnected* is the operative word that distinguishes cognitivism from associationist theory and the classical illustration of animal cognition, cited by Cheney & Seyfarth (ibid., p. 8), is Wolfgang Kohler's conclusion that his captive chimpanzees were showing insight when they suddenly started to join sticks or pile up boxes to reach food. It has since been recognized, however, that this was not strictly a new use of the sticks and boxes. The animals had to have had previous experience of putting these objects together if they were to do so to get food (see McFarland 1981, pp. 309 and 313, or 1985, p. 349).

Animal cognitivism thus seems to be another overswing of the theoretical pendulum, this time back towards anthropomorphism.

> "out of the disillusionment [with radical behaviourism] the hydra-headed monster of mentalism, once thought to have been subdued by Watson and finally despatched by Skinner and Ryle, has resurfaced in the form of contemporary cognitivism." (Lowe 1983, p. 73)
>
> "What many attempts to define cognition seem to hint at is that we should regard as evidence of cognition anything remotely akin to human cognition. This is...merely a covert way of adopting an anthropomorphic posture." (McFarland 1989*a*, p. 132)
>
> "cognitive ethology is no advance over the anecdotalism and anthropomorphism which characterized interest in animal behaviour a century ago." (Colgan 1989, p. 67)

But neobehaviourist cognitivists assure us that they are materialists (Wilder 1990; Bekoff & Jamieson 1991) and Terrace underlined the point, saying "Skinner should be heartened by these...demonstrations of the feasibility of studying complex processes in humans and animals from a monistic and a materialistic point of view" (Terrace 1984*c*). Not least because of such reassurances, as Dunbar (1989) rightly said, "traditional ethology notwithstanding, cognitive ethology has now become a boom industry". It is driven by a feeling that "at present, one of the most exciting areas of research in the behavioural sciences lies in the interface between linguistics, computer science, cognitive psychology, and philosophy" (Cheney *et al.* 1987). Burghardt is an enthusiastic animal cognitivist who nevertheless takes a very realistic view of the problems facing himself and others whose object is "direct scientific study of the animal mind" (Burghardt 1985, p. 917*a*). He is well aware of the Scylla and Charybdis between which they must steer: "Unless challenged to separate description from interpretation, students readily use and defend the use of sloppy teleological and anthropomorphic *thinking*, not just the vocabulary. On the other hand, a studious, ideologically based opposition to using our own experience and intuitions... in asking questions and designing studies in animal behaviour, is ultimately sterile and dull" (ibid., p. 917*a*). But Burghardt's main anxiety is that his fellow cognitivists are "in danger of repeating the failure of early supporters of animal mentalism even if they never refer to consciousness or awareness" (ibid., p. 908*b*), and thus of playing into the hands of the behaviourists again. His reason for this anxiety is that he thinks modern cognitive psychologists and ethologists have not, so far, made any real advance beyond the position of psychologists at the end of the last century. "It is only fair to ask what exactly is new here that was not only foreshadowed but also developed more fully by Romanes and Morgan" (ibid., p. 913*b*). Another cognitive psychologist, Wasserman, has said much the same: "One can question Griffin's... conviction that the prospects of understanding the subjective mental experiences of animals – their sensations, feelings and intentions – is any better now than in the days of Darwin and Romanes" (Wasserman 1984). When Romanes and Morgan failed in their endeavour to illuminate animal minds, Burghardt said that they left a void that was filled by behaviourism: "Loeb (1918), Washburn (1908), Jennings (1906) and Holmes (1911) all, for differing reasons, became skeptical of gaining verifiable information from introspective analogies" (Burghardt 1985, p. 913*b*).

Burghardt (1985) acknowledged that there were some new resources at the disposal of modern animal cognitivists: "newer models, often computer based, from information processing, psycholinguistics, economics, and so on" (ibid., p. 908*b*), as well as the work on self-awareness in apes (**5.3**) and semantic communication in apes (**3.4**) and vervet monkeys. Nevertheless, he found that

> "mentalistic thinking has, at best, shaped the research done [by animal cognitivists] but has not helped us gain direct knowledge of what the monkeys experience.... What we seem to have returned to is the hoary issue of mind–body dualism.... The widely recognized central issue is how to proceed from our [introspective] 'experience' to another animal's" (Burghardt 1985, p. 915*b*)

He looked forward to more success in the future but his design for the future seemed to be a rather forlorn rallying cry: "It is time that animal workers of all persuasions join with developmental, physiological and cognitive human psychologists to address the entire issue of mentalism in psychology.... Until that time let us not think just about semantic issues in mentalistic terminology or how to draw lines as to what animals reach what abstract level. Let us stick to our last, refer to specific abilities (e.g. selective attention, communication of resource location, mirror self-recognition), and keep the referents of our concepts to the fore. Let us retain an open-minded delight in animal abilities, a respect for what they *may* be experiencing, and a balance between skepticism and incredulity" (Burghardt 1985, p. 918*a*). Such unquenchable faith in the eventual demonstration of animal cognition speaks eloquently for our in-built empathy towards animals. The cognitive-ethological philosopher Wilder (1990) showed a similarly unshakeable commitment. On the one hand, he described in uncompromising detail the manifold and still continuing risks of Clever Hans errors that beset the ape language work (**3.4**). But he then declared his continuing belief that "cognitive science is a possible science" simply because grounds for scepticism remain (ibid., pp. 364–5). An astrologist or phrenologist could happily use that breath-taking argument. In science it is not usual to lean so very far over backwards to give hypotheses the benefit of the doubt. There seems to be some special incentive for the unusual indulgence shown towards the cognitivist hypothesis here and, presumably, it is unwitting anthropomorphism.

Griffin (1984, p. vi) as an explicit anthropomorphist criticized modern animal cognitivists for leaving out consciousness. He said he used "*cognitive* in a literal sense to refer to conscious thought and knowledge, thus avoiding a recent tendency to restrict the term to information processing as in cognitive psychology". Many contemporary animal cognitivists follow Griffin there (e.g. Mellgren 1983; Bekoff & Jamieson 1990*b*). But for many others, with a more neobehaviourist outlook, "consciousness ... has not been a major issue in the recent revival of interest in animal cognition .... The rationale for the study of cognitive processes in animals requires no reference to animal consciousness. Both in human and in animal cognition it is assumed that the normal state of affairs is unconscious activity and thought" (Terrace 1984*b*, p. 8; cp. Cheney & Seyfarth 1990). At first sight there may seem to be no anthropomorphism there and indeed there is none in conscious, explicit form (although "unconscious thought" will seem paradoxical to many). But disavowing animal consciousness is not a complete rebuttal of the charge of anthropomorphism as Burghardt and Terrace seem to believe. Terrace took Skinner to task for "his reluctance to acknowledge that the study of representations does not imply a regression to mentalism" (Terrace 1984*c*). But Skinner's reluctance did have some justification: see the quotation from Bekoff & Jamieson (1991) on p. 98. (In ordinary use cognition may be considered to be conscious as Griffin says, but anthropomorphism is implicit also in the cognitivist practice of taking it for granted that internal representations of one kind or another play a guiding role in many animal behaviours. Such control may occur, but six instances have already been given where simply assuming that it occurs has led to a mistake (**3.2**, **3.3**, **4.1**, **4.2**, **4.3** and **4.4**).

The mistakes generated specifically by cognitive interpretations are distinguished by what Staddon (1989) called needless "complexification". As an example he cited a simple theory of the timing behaviour of rats "which has all the complexity of an egg timer" yet "is for some reason termed 'cognitive'". Colgan gave two more: "Even operant conditioning of stickleback responses ... and learning of foraging responses in Atlantic salmon ... are dignified as examples of cognition when in fact simple accounts in terms of familiar, basic mechanisms are sufficient for explanation" (Colgan 1989, p. 65). A striking example of this "complexification" comes in Toates's (1986) book *Motivational Systems*, already cited on another matter on p. 69. Espousing "a cognitive, purpose model of motivation" (ibid., p. 11) he gave an illustration of what he meant by cognition. He contrasted the performances of rats trained in two different ways to find food in a T- or Y-maze (ibid., pp. 9–11 and 13–14). In the first case the rat was trained to find the food by turning left at the junction. In the second case the rat was trained to find the food by turning to the side on which it could see a window outside the maze, whether that was on the left or the right. In the second case only, Toates says the rat "learned where food is in a form not tied directly to behaviour". Instead, the training created in the rat an "expectation" that the food was "over there by the window", and this, "through cognitive processes, shaped its behaviour". There is no apparent reason for making so great a contrast between the two performances. Cognitive psychologists have differentiated between "response learning" and "place learning" (Restle 1957), terms which they might consider to be appropriate here, but the fact is that the two performances differed only in the nature of the learned cue to turn one way as opposed to the other: in the first case, close-range tactile or proprioceptive stimuli received at the runway junction; in the second case, a distant visual cue visible from the runway. There seems to be no more reason for invoking the complex processes of expectation and cognition (including the topographical concept of "over there") in the second case than in the first.

Another instance of "complexification" through invoking cognition, more interesting because less obvious, occurs in the same book: "Klinger demonstrates the need for a cognitive construct in the following comparison: 'A mosquito wafted out of sensory contact with a warm body is unlikely to retrace its steps in order to take up where it left off, but a dog that has lost sight of a thrown stick may search for a while'" (Toates 1986, p. 15). Now it has been known for many years that insects and dogs behave quite similarly in such a situation. A mosquito, or one of sundry other flying insects, will fly up a warm, odorous or otherwise stimulating air stream emanating as a plume from some valuable resource object such as a host or mate to the windward. If the insect loses contact with the plume somewhere along the way, it may thereupon resort to much the same casting about, dropping back and re-orientation manoeuvres as a dog does when losing a scent (Wright 1958, 1962, 1964; Daykin *et al.* 1965; Traynier 1968; Brady *et al.* 1989; Baker 1990). Insects of course excite our empathy less than do dogs and it has not occurred to modern students of insects to credit them with cognition because of that kind of performance. If Klinger and Toates felt sure that mosquitoes were unlikely to behave in that way, like a dog, this could only be because they supposed that mosquitoes lacked the cognitive ability that they believed was needed in such a case and was possessed only by more complex animals like dogs. If they had not assumed, anthropomorphically, that the dog's performance depended on cognition they would not have unwittingly underestimated the behavioural capacities of mosquitoes.

Note. It is because of the anthropomorphic content of cognitivism in general that I have ventured to purloin the term neobehaviourism for use in the novel sense defined in Chapter 1. This sense is quite different from the traditional sense of neobehaviourism that is discussed at length by Amsel (1989). He makes a sharp distinction between (a) the neobehaviourists in the traditional sense such as Tolman, Hull and their present-day descendants such as Amsel himself, whose outlook originated with that of the early Watson before he hardened into a radical behaviourist, and (b) today's animal cognitivists such as Roitblat (1987), Burghardt (1985) and others cited above who are "cognitivists in a much more polarized sense" (Amsel 1989, p. 1). Amsel maintained that the former authors employ cognitive concepts only in "explaining behaviour", whereas the latter use them in "studying the mind" (ibid., p. 25). Using his own position to illustrate that traditional meaning of neobehaviourism, Amsel (ibid., p. 2) wrote "Although... I work with and theorize about mechanisms emphasizing mediation and anticipation, that can obviously be characterized as cognitive, my empirical constructs are still clothed in the stimulus–response language associated with behaviorism. I find this approach at the same time more analytical and more constraining than the more mentalistic cognitive approach."

That seems rather a difficult balancing act to maintain; but some combination of behaviourist and cognitivist concepts (more or less unwittingly mentalistic) is not uncommon today among (vertebrate) zoologists and psychologists alike. Thus Krebs (1977) once remarked in passing that "most behavioural ecologists are quite ready to accept the idea that animals have mental images of events or objects in their environment". Cognitivism is more usual among psychologists but from the anti-anthropomorphist viewpoint the many differences among the people that Amsel discussed under (a) and (b) above are less important than the cognitive element which unites them. They also share the anti-anthropomorphic legacy of radical behaviourism while at the same time rejecting the school's oversimplifications. 'Neobehaviourist' then seems a very convenient name to embrace them all.

5.3 Self-awareness

Claims that a chimpanzee could learn to use grammatical human language expressed in the form of visual symbols were once very widely accepted but recognized as mistaken within a few years (**3.4**). On the other hand, another, even more ambitious anthropomorphic claim, that a chimpanzee can become 'self-aware', was put forward at about the same time by Gallup (1970, 1971, 1977), and this has won wide acceptance which has not been withdrawn subsequently (Slobodkin 1977, p. 337; Groves 1978; Crook 1980, p. 243; M. Dawkins 1980, p. 16; Gould 1982, p. 483; Griffin 1984, p. 30; Anderson 1984; Passingham 1982, pp. 50 and 239; Humphrey 1986, p. 82; Goodall 1986, p. 35; Mitchell, 1986). Certainly, the experimental evidence on which this claim is based is striking and extensive.

First, four young chimpanzees were exposed individually to full-length mirrors for eight hours a day for ten days. Initially the chimps gave only 'other-directed' responses to the mirror while watching their reflected images; that is to say they acted as if the image were another chimp, bobbing, vocalizing, 'threatening', etc. After about the third day these other-directed responses had waned and instead the chimps made 'self-directed' responses to their mirror-images, inspecting, grooming and manipulating parts of their bodies which they had never seen before, posturing, looking at themselves upside-down, blowing bubbles, making faces, picking bits of food from between their teeth, removing extraneous material from the corners of their eyes and nose. They kept their eyes glued to the mirror while putting on all this performance. The time they spent looking in the mirror peaked on the third day, then declined. In the next test the mirror was removed after the tenth day, and while the chimps were anaesthetized two bright red marks were painted on their faces (where the chimps could see them only in a mirror), using an odourless, non-irritating, alcohol-soluble dye. On recovering from the anaesthetic the chimps showed no after-effects until the mirror was restored to them, but then the time they spent looking in the mirror sharply increased again. They repeatedly touched the dye-marks while watching their reflections, and looked and sniffed at their finger tips after touching a mark. Similar dye-marked chimps that had had no previous mirror experience made only other-directed responses on seeing themselves in a mirror.

These results, obtained with chimpanzees and orang-utans, have been amply confirmed. At the same time tests with a great variety of other primates including another great ape, the gorilla, have consistently failed to yield any evidence of a capacity to learn to make 'self-directed' responses to their own mirror-image, with and without dye-marks. Some of these tests with monkeys were kept going for years, but in vain. Yet a monkey is perfectly capable of using a mirror as an optical tool to manipulate objects it cannot otherwise see and can even learn to turn away from a mirror it had been confronting and thereby gain closer access to an object behind it that it had first seen in the mirror (Anderson 1984).

From these results Gallup (1977) concluded that what a monkey lacked and could never learn to acquire was "a sense of self". Whereas a chimpanzee which had learnt to respond to its mirror-image 'self-directedly' must have "recognized" itself and was therefore "self-aware". The core of the argument was reiterated in several publications as follows:

> "One of the unique features of mirrors is that the identity of an observer and his reflection in a mirror are necessarily one and the same. Therefore, if you do not know who you are, how could you possibly know who it is you are seeing when confronted with your mirror image? The capacity to correctly infer the identity of the reflection presupposes an identity on the part of the organism making that inference. The monkey's tendency to continue to react to himself in a mirror as if he were seeing another monkey may be due to the absence of a sufficiently well integrated self-concept." (Gallup 1982, p. 240; 1983, p. 484; 1987, p. 4)

What one does not notice at first reading of that passage is that the writing changes gear, so to speak, without warning. Up to that point we had a straightforwardly objective account of what was observed, but then in that passage there is a sudden, easily overlooked resort to analogy, jumping from chimp to human being. This shift is evident from the "you ... you ... you ... you ... your ..." in the second sentence. This implicitly asks us to take it for granted that what is true for the human is true for the chimp. The next sentence carries a like assumption: that a chimpanzee can make inferences as we do. In both cases the question being asked 'are chimpanzees self-aware?' is unwittingly begged. The making of such unsupported assumptions reveals a subconscious anthropomorphic bias.

Virtually that same criticism of Gallup's interpretation of his work was levelled by two commentators on Griffin's (1978) article, which attached great importance to Gallup's work: "Of course, if we are required to describe the animals in Gallup's experiment as Griffin does, viz. that 'they recognized the mirror image as a representation of their own bodies,' then we do seem to be committed to postulating that the animals have the connected conceptual powers. But why... should we not say instead, for example, that an animal has learned to react to the mirror input by treating it as a set of cues for reacting to the corresponding part of its own body? That description, or something much better,... does not commit us to ascribe conceptual and propositional powers to the animal" (Farrell 1978). The other commentator made a similar and clearer proposal: "that a mirror-educated chimpanzee immediately rubs off a spot on his forehead when he sees it in a mirror is not... 'clear evidence for self-awareness,' at least in its usual sense.... Our conscious selves are not our bodies... we do not see our conscious selves in mirrors. Gallup's chimpanzee has learnt a point-to-point relation between a mirror image and his body, wonderful as that is" (Jaynes 1978). Jaynes's interpretation, in effect that the chimp forms a point-to-point association between the movements of the mirror image and his own movements, would enable the chimp to do what the monkey never learns to do, use the mirror as an optical tool for manipulating his own body. Every movement will provide his brain with proprioceptive and tactile feedback from his own body directly, accompanied by simultaneous visual feedback from the mirror – the kind of association he is familiar with when he manipulates parts of his body that he can see and monitor directly. His persistent and multifarious 'performing' while watching his mirror-image will provide his brain with countless repetitions of the concurrence of those two kinds of sensory input, one from his body and the other from the image. This could ensure that their association was thoroughly learnt: a kind of 'auto-shaping'. Once the association was formed, it could be used for visual guidance of actions that could not be followed by sight directly.

What, then, would be the feature of chimpanzee behaviour that is missing in most primates but enables chimpanzees to learn to respond self-directedly to a mirror? It could be that the chimpanzee (and orang-utan) alone responds to the stimulus of synchronous input of those two kinds by engaging in a great deal of motor activity ('performing'), and thereby generates more and more of such synchronous concurrences while it is watching its image closely. In other words, when a chimp's movements cause an imaged chimp in a mirror to duplicate those movements instantly and continuously, the movements may be self-reinforcing, producing positive feedback. If a monkey did not respond in that way to duplication of its movements in a mirror, it could not learn the association between its own movements and those of its image and use it for visually guided manipulations of its own body. This hypothesis or something like it is plausible enough to need ruling out carefully before we are asked to accept any more ambitious interpretation of the results. Unfortunately it was not ruled out before the 'self-concept' interpretation was confidently advanced and widely accepted. Like the too-ready acceptance of the 'search image' hypothesis (see **3.3**), this omission again shows the influence of an unwitting anthropomorphic bias.

As a matter of fact, Gallup himself came near to formulating the above associationist hypothesis of the self-directed behaviour of mirror-trained chimps when explaining his definition of self-awareness, which was, "an organism is self-aware to the extent that it can be shown to be capable of becoming the object of its own attention" (Gallup 1987, p. 3). When grooming itself "the organism is ostensibly directing its attention to parts of itself [but] this does not presuppose that it is able to conceive of itself as a separate, independent entity with an identity of its own. When a monkey grooms its own arm, the monkey is not the subject of its own attention; its arm is. Indeed, I would argue that because of proprioceptive feedback and kinasthetic cues, coupled with response-contingent stimulation, the appropriate use of your arm *does not require that you know it is yours*" (Gallup 1987, p. 3; my italics). We have only to consider the situation where it is the chimpanzee's own image in a mirror that provides the "coupled response-contingent stimulation" of which Gallup spoke, and we have there the alternative, associationist mechanism that I have proposed above for the self-directed behaviour of a mirror-trained chimp. This hypothesis could be tested by comparing a chimp's reaction to 'live' (closed-circuit) TV pictures of itself with a videorecorded picture of itself in the same situation. The recorded picture would contain all the individual characteristics that would enter into a pictorial "self-concept", except one. It would lack only the instant duplication of the subject's own movements that he sees in the 'live' TV picture or a mirror. Initially naive chimpanzees could be presented with 'live' and taped TV pictures of themselves alternately every several minutes for an hour or so a day, day after day. The hypothesis would be confirmed if they began to respond self-directedly to the 'live' image before they did so to the taped one. Given the same treatment for enough further time to learn the individual characteristics of their image (after they had already learned to respond to the live one self-directedly) they might eventually come to respond self-directedly to the recorded picture as well, but these responses would be expected to be fewer and briefer than to a live picture.

As it happens both live and taped TV pictures of themselves have been presented to chimpanzees by Savage-Rumbaugh (1986). Her account confirms Gallup's results with mirrors but it is anecdotal and unfortunately does not include any systematic experimental comparison of responses to the two types of image as proposed. She wrote: "Sherman and Austin [*the chimpanzees*]... came to differentiate live from taped portrayals of themselves spontaneously by testing the video image when they saw themselves on the screen. That is, they made faces, stuck out their tongues, waved their hands and feet, looked at their mouths, first on the TV monitor, then in the mirror, then on the monitor, then in the mirror, etc., groomed themselves as they watched the monitor, and even pretended to feed themselves as they watched their own images on the monitor, not when they saw the images of other apes.... Once they ascertained, through such testing, that they were viewing taped images of themselves, they typically showed little interest in the monitor. By contrast, when they determined that their images were live, they often continued to 'play' with the image for some time" (Savage-Rumbaugh 1986, pp. 309–10). Thus as far as one can tell it seems that after an unknown amount of prior experience with TV images the chimps did show some 'performing' responses to their own taped image specifically, as opposed to that of another ape. As expected if the concurrence of their own and the image's movements is the main requirement for 'performing' and then responding self-directedly, the responses to the taped image were apparently brief and transient whereas those to the live image were intense and prolonged, again judging by the anecdotal account.

More decisively, Menzel *et al.* (1985) confirmed the postulated requirement for concurrence of the chimp's own movements of his image, for self-directed responses to occur. For this work, they did not use images of whole chimps. Instead they compared a chimp's responses to 'live' video images of his own hand and arm poking through a hole in a panel, with his responses to taped video images of the same scene. These images were presented simultaneously on separate monitors on the chimp's side of the panel. There was some target object, say a raisin, fixed to the side of the panel away from the chimp and not visible to him directly but visible to him in both monitors. The chimp has already learned to use a live TV picture of its hand to guide it to a target object seen in the same picture but not directly visible to the chimp. The taped image in the experiment was of the same scene at some previous time and could not be distinguished from the live image until the animals 'tested' it by sticking a hand through the hole and waving it about, movements which of course were not duplicated by the hand seen in the taped image. "The chimpanzees began within two or three trials to stick their hand out 'into the picture' only for an instant and then to either proceed as usual [reaching and contacting the target] (if they had chanced to pick the live image) or turn immediately to the other monitor (if they had picked the tape)" (Menzel *et al.* 1985, pp. 214–15).

To sum up, it would of course be rash to say that no animal

A chimpanzee using closed-circuit TV to guide its hand to an object that it cannot see directly. (Reproduced from Menzel, E. W., Savage-Rumbaugh, E. S. & Lawson, J. 1985. *Journal of Comparative Psychology* **99**, 211–17. Copyright 1985 by the American Psychological Association. Reprinted by permission.)

will ever be shown to be self-aware; Gallup's self-awareness hypothesis has not been disproved. It is hard to see how it could proved or disproved. But for present purposes what matters is that his hypothesis has been accepted without ruling out a more plausible alternative hypothesis, and this indicates an anthropomorphic bias, even if it should turn out that the alternative hypothesis is wrong, which of course it could well be.

Note how the descriptions of the behaviour of the chimps in those two sets of experiments with TV images display an unconscious defect in much primatological writing. This is the appearance of out-of-place, anthropomorphic formulations incorporated smoothly and unhesitatingly into otherwise objective accounts of the animals' behaviour, as in the passage from Gallup quoted at the beginning of this section (p. 107). This is not 'mock' anthropomorphism (pp. 88–9) but the real thing, although unwitting. Note likewise "testing" and "ascertained' in the passage from Savage-Rumbaugh (1986) above, and "behavioural awareness", "realized", "understood", "intrigue" and 'knew" in the following sentences:

> "Sherman and Austin's intrigue with the video portrayal of the [food-selecting] activities of one another and the teacher in the next room, coupled with the overt response of running into the adjacent room when given permission to retrieve the food, clearly revealed that they knew what it was they were seeing on the monitor." (Ibid., p. 306)
>
> "We have an excellent record of the first occasion on which Austin evidenced behavioral awareness that the chimpanzee on the TV screen was himself.... As he glanced over at the TV monitor he suddenly realized that he could see the orange drink in his mouth better than by looking down over his nose." (Ibid., p. 308)

5.4 Suffering

"The conviction that it is possible to draw an analogy between suffering in ourselves and that in other species is probably the basis for people's concern about animal welfare" (M. S. Dawkins 1980, p. 111). People perceive the analogy as the logical basis of their concern if they stop to think, but they feel the concern first. In our culture, the sight of a higher vertebrate, especially a mammal, suffering an injury, or of an uninjured one in circumstances that look to us very unpleasant, arouses in most people an instant feeling of pity for the animal, just as does the sight of a fellow human being in like straits. The feeling is spontaneous and unthinking. It is empathy, and it underpins our intuitive belief that higher animals have minds as we do (see **2.6**, **2.7**, **3.4** and **5.2**). That is to say the layman's approach to the problem of animal welfare is unhesitatingly anthropomorphic. There is no avoiding the fact that the public demand for legal protection of animals is fuelled by this feeling more than by logic. But, after all, the public includes scientists and not least neobehaviourists. They too feel spontaneous sympathy for animals that they feel are ill used, although as scientists they have no adequate evidence that the animals do actually suffer. We cannot escape from this inner contradiction. Scientists really are in a dilemma; they do not merely appear to be in one as M. S. Dawkins (1990) argued.

Van Rooijen is an ethologist working on farm animal welfare who strongly supports the behaviourist view that it is unscientific to assume that animals have feelings. Nevertheless, he confessed that in practice he did assume that animals had them, "the behaviour of one maltreated animal" being reason enough: "we do not need other arguments" (van Rooijen 1981). In other words van Rooijen recognized that there were two contradictory strands in his approach to the subject, one based on scientific reasoning and the other on a spontaneous feeling. Blaise Pascal's famous aphorism would not be out of place here: "the heart has its reasons that reason knows nothing of." Van Rooijen thought it was futile to try to reconcile these two contradictory attitudes and felt it was his responsibility to admit this so as to preserve the scientific reputation of applied ethology. This would seem to be a realistic position to adopt. M. S. Dawkins (1990, p. viii) thought she might be accused of 'sitting on the fence' but van Rooijen sensed that what scientists are sitting on here in fact are the horns of a dilemma. However, most ethologists think quite otherwise. Thus M. S. Dawkins (1990) believes "a middle way is possible. We can acknowledge the genuine difficulty of ascertaining what a nonhuman animal feels and yet attempt to attain a scientific understanding of its feelings." In the same vein Sperlinger's (1981) Introduction to a multi-authored book on *Animals in research* ended as follows (ibid., p. 70): "this book ... will ... have served its purpose if it ... brings forward recognition of the fact that 'animal welfare' and 'science' are not necessarily irreconcilably opposed". The traditional conflict that Sperlinger was referring to there was not of course the same as van Rooijen's conflict between feeling and reason within one person. But in both cases the conflict is between 'heart' and 'brain', which are irreconcilable in principle, because different in quality. We can attain an empathic 'understanding' of an animal's feelings, but not a scientific understanding of them; we have no scientific knowledge that it has any.

At the conclusion of a 1980 interdisciplinary conference on animal awareness in the context of animal welfare: "all delegates, sceptics, agnostics and protagonistics [*sic*] agreed that no single experiment or set of experiments could conclusively prove the existence of self-awareness in animals below the Great Apes" (Wood-Gush *et al.* 1981). In her book *Animal suffering* M. S. Dawkins was equally definite on this point, but eventually somehow seemed to retreat from it and concede the scientific legitimacy of arguing by analogy from ourselves, in this passage: "it would be unfortunate if this book had left the impression that finding out about physiology and behaviour was enough to tell us whether animals suffered ... no amount of measurements can tell us what animals are actually experiencing. Their private mental experiences, if they have them, remain inaccessible to direct observation .... So, if we are to conclude that animals do experience suffering in ways somewhat similar to ourselves, then this conclusion has to be based, in the end, on an analogy with our own feelings" (M. S. Dawkins 1980, p. 102). "Such analogies should, however, be drawn only with the help of all possible evidence about the animals concerned" (M. S. Dawkins ibid., 111–12). So, *given* that help, we are to infer that we can, after all, draw the analogy.

M. S. Dawkins's (1980) book is, as Rollin testified, "the main apologia for reintroducing these banished notions [of 'negative subjective experiences in animals'] into science .... The core of her book consists of a discussion of the various methods by which one can assess the suffering of animals. Though no single method provides indubitable proof, taken together these methods justify the attribution of suffering or happiness' (Rollin 1989, p. 256). M. S. Dawkins, like most ethologists but unlike van Rooijen, is sure that we can use other arguments to strengthen the case for the existence of animal suffering. "When all the evidence has been collected, we have to return again to analogies with ourselves for the final verdict ... However, ... they are very different analogies ... disciplined by scientific method and informed about the biological needs of the animals ... far more likely to yield an accurate picture of what an animal is feeling than uninformed analogies ever could " (M. S. Dawkins 1980, p. 106). "Even if mental phenomena cannot be studied in exactly the same ways as behaviour and physiology ... very strong indirect evidence can be accumulated about them" (ibid., p. 23). "The point of this book is not to eliminate subjective judgements altogether from our analysis of whether animals are suffering. The point has been to put those subjective judgements on a scientific footing..." (ibid., p. 116).

The 'take-home' impression with which we are left, then, is that, although there is no direct evidence that animals suffer the indirect evidence is so strong that the proposition can be accepted. This is likewise the view of Bateson (1992). However, the indirect, scientific evidence cannot be good enough to permit us to say our subjective feeling that animals suffer now rests on a scientific footing. The objective and subjective evidence are different in kind, 'immiscible'. And the objective evidence itself is by no means straightforward. Even the seemingly elementary first step of describing the neurosensory inputs from wounding has become anything but straightforward. "Gone is the search for 'labelled line' nociceptive pathways.... The specific central nociceptive neurone has been replaced by an altogether more plastic nociceptive neurone subject to changes in excitability and long-term alterations in the effectiveness of synaptic inputs; a neurone may respond to noxious and/or non-noxious inputs depending on the circumstances" (Fitzgerald 1990). We cannot reliably infer pain sensations from non-verbal behaviour even in human beings. Among "patients admitted to hospital with severe injuries, 40% noticed no pain at the time of the injury of which they were fully aware, 40% had more pain than expected and only 20% reported the expected pain" (Wall 1985). Similarly, Bowsher, in his review of "Pain sensations and pain reactions" at the 1980 conference already mentioned, said:

> "A person suffering an injury on the battlefield or sports field may be almost completely unaware of the pain." (Ibid., 1980, p. 25.) "*Decerebrated animals... are capable of an*

> *integrated reaction to noxious stimuli, including... vocalization*.... Human beings to whom minimal doses of short-acting anaesthetics are administered for the purpose of dental extraction... may jump, writhe, and cry out... and yet subsequently are so unaware that a tooth has been extracted that they frequently ask when it is to be done. Thus we must be careful to distinguish, in man as well as other animals, between pain sensation and reaction to a painful stimulus, noting carefully that the two can be totally dissociated." (Ibid., p. 23.) "There is a long way to go yet before we understand the mechanisms of pain and suffering in man... thus we are even further from comprehending the relationship between stimulation and perception, and perception and consciousness, in other species." (Ibid., p. 27; my italics)

None of this proves that animals do not suffer, of course. What it does show is that we cannot hope to tell from their behaviour whether they suffer or not.

Overreadiness to believe that animal suffering has been as good as proved shows that there is an unwitting anthropomorphic bias, stemming from the spontaneous fellow-feeling for higher animals which scientists have in common with other people. For there is "a primitive and almost irresistible tendency" to attribute human mental states to pets and other animals (Gallup 1982, p. 243; cp. Hediger in **3.4**). This anthropomorphic bias is often apparent in scientists' approach to animal welfare although *suffering* was actually described as "a subjective term if ever there was one" in the report of one of the working groups at an elite international (Dahlem) conference on animal mentality in 1981 (M. S. Dawkins 1982, p. 371). Since it is indeed a subjective term, book and chapter titles such as 'The physiology of suffering' imply that we already know that animals do suffer, thus begging the question to be discussed. For example, "The growing concern for animals... is a concern that some of the ways in which humans treat other animals cause mental suffering and that these animals experience 'pain', 'boredom', 'hunger', and other unpleasant states perhaps not totally unlike those we experience" (M. S. Dawkins 1990, p. 1). The phrase "perhaps not totally unlike" displays scientific caution and so do the inverted commas around 'pain' and other terms. But at the same time *mental suffering* and *unpleasant*, although equally subjective, are left free of inverted commas. This again begs the question of whether animals suffer. To define suffering as experiencing unpleasantness is tautological, or "going from one subjectivity to another", to borrow the title of Shettleworth & Mrosovsky's (1990) commentary on M. S. Dawkins (1990).

M. S. Dawkins has now introduced a powerful technique, borrowed from economics and known as 'demand curve analysis', to measure how much an animal is prepared to do to avoid something or to obtain something. "If an animal, even one totally different from myself, shows evidence of clear-cut behavioural priorities as revealed by extensive demand curve analysis, if the animal appears to be prepared to do almost anything to obtain something even when it is made difficult to do so, if the animal will learn an operant response to get something and shows evidence of behavioural and physiological changes when deprived of it – if the animal does all these things, then this would for me constitute powerful evidence of a capacity to suffer" (M. S. Dawkins 1990, p. 53). Demand curve analysis certainly has the advantage that it "frees us from an entirely human-centred view of animals" but that is not to say it is tantamount to "asking the animals themselves" or that it brings us "close... to the animal's point of view" (ibid., p. 54). Application of this technique is to be welcomed as it is likely to be widely accepted as a rationale for outlawing any treatment of animals that is abhorrent to us. But let us have no illusions; it will leave us indefinitely far from "the animal's point of view", since that is only a metaphor, as is "asking" the animals. It will not in fact provide us with "a bridge between our subjective world and theirs" (ibid., p. 4), and to believe that it will do so is to beg the question by taking it for granted, with unwitting anthropomorphism, that the animals have a subjective world. Sadly, there does not yet exist a "scientific study of animals suffering" that "can give us an insight into what animals experience" (ibid., p. vii). Toates (1990) took the only tenable position for a scientist in this matter today when he said, "The question of whether animals suffer must remain unresolved."

In his recent substantial treatise on animal consciousness and suffering Rollin (1989) pointed out that scientists sometimes come up with conclusions that everyone has long taken as common sense, and which, he argued, the scientists should never have written off. He claimed that this is what has happened recently in the scientific rehabilitation of subjectivist concepts of animal consciousness, suffering and cognition. Goodall's Foreword to his book sums it up: "...the common-sense view of most people has always been that animals...experience a variety of human-like feelings including pain. It was from this perspective that Darwin, for example, argued that there was continuity in the evolution of the mind as well as structure. Subsequently, however, American psychology introduced the concept of behaviourism and it became fashionable to view animals as animated, mindless machines – walking bundles of stimulus-responses.... It is the growing moral concern for animals and their welfare among the general public that is putting pressure on scientists to investigate animal consciousness and suffering" (Goodall 1989). Rollin, like other protagonists of animal rights (e.g. Singer 1985), takes it for granted that the common-sense view that animal consciousness is real, is correct and no proof is needed. Indeed it does appear to be correct if we assume that animals without consciousness would be mere "walking bundles of stimulus-responses" – but that is a travesty of the neobehaviourist's conception of animals, a travesty to which explicit anthropomorphists seem to cling (pp. 3 and 62–3). It might seem necessary to suppose that some animals have minds if we had no other explanation for their flexible, adaptive behaviour. But there is of course another explanation, namely the power of natural selection to optimize behaviour along with the other features of organisms (**5.1**). But anthropomorphic empathy towards animals is evidently sometimes strong enough to brush that scientific point aside. Rollin and Singer took their anthropomorphism still further in insisting that we have a moral obligation towards animals, claiming they have rights as do the citizens of some human societies. Morality is of course a creation strictly of human society and it is merely an empathically inspired fantasy to suppose we can grant animals the status of honorary humans for this purpose.

The anthropomorphic bias in the case of animal suffering is obviously not just a matter for scientists but impinges on the public domain. "Just because we are professionals in a cognate discipline, we do not own this issue in the same sense that we can expect to give the last word on questions well within our fields of special expertise. A large segment of the public knows what it thinks about the mental life of familiar animals ... hence laws and societies to prevent abuse" (Bullock 1982). "The public's view on the use of animals as research subjects has moved noticeably in the direction of giving greater rights and attention to animals. Insofar as professional scientists have tried to oppose this trend, they have merely reinforced a growing public distaste for science" (Driscoll & Bateson 1988).

It is incumbent on minorities in a democracy to respect laws that enshrine the perceptions of the majority, be these perceptions true or false. Scientists who think that there is inadequate evidence that animals suffer constitute such a minority; in any case that is a conclusion they reach with their "heads" while their "hearts" are mostly with the majority. "Desirable as a more rational approach may be, we doubt that it [*demand curve analysis*] will readily replace the compassionate and generally informed, but ultimately subjective, assessment of what treatment of animals is permissible" (Shettleworth & Mrosovsky 1990).

Of course one cannot rule out the possibility that one day someone will devise a satisfactory test for feeling in animals. But there is none in sight and meanwhile it requires an anthropomorphic bias to be convinced they have feelings. Scientists have therefore to live with an internal conflict. If they encourage people to assume that there are good scientific reasons for believing that animals do experience suffering in some degree, knowing that such reasons do not exist, that will do science (and animals) no good in the long run. Scientists' reputation for respecting the truth is something too precious to them, and to society at large, to be jeopardized in that way. Thus, Lockwood (1985/6, p. 198) quoted C. W. Hume's statement as the clinching last word on the subject: "If I assume that animals have feelings of pain, fear, hunger and the like, and if I am mistaken in doing so, no harm will have been done; but if I assume the contrary, when in fact animals do have such feelings, then I open the way to unlimited cruelties," is not quite as self-evident as it sounds. For if scientists say they have evidence that animals have subjective feelings such as pain although they do not have this evidence, some harm will be done – to science; if they assume that animals do not feel pain, or have doubts about it, they can nevertheless quite sincerely support legal measures and professional codes of conduct (e.g. Anon. 1991) to prevent animals being treated in ways that most people believe are unkind, because such treatments arouse feelings of pity and revulsion in most scientists too. To give such support is to give the animals the benefit of the doubt in the spirit of Hume. We have an additional motive for giving it, adumbrated by Immanuel Kant according to Midgley (1985), inasmuch as treatment of animals which is seen by whoever carries it out as callous may encourage them to treat people callously too, especially people they see as in some way sub-human (Diamond 1981). Fox (1986) judged this to be not literally true, but at the same time he elaborated five other reasons for treating animals humanely, one of which is the demeaning effect that maltreating them has on people who do it.

CHAPTER 6

The three case-studies making up this chapter deal with unwitting anthropomorphism in forms not already discussed.

6.1 Hierarchy

In **3.4**, unintended anthropomorphism was held to be partly responsible for the eventual disappointment of early hopes of bridging the great gap in knowledge between "what nerve cells do and how animals behave". But anthropomorphism has served indirectly to frustrate those bridge-building hopes in another way, by compromising the important principle of a hierarchy among the causal mechanisms of behaviour. It was probably Tinbergen (1950, 1911) who brought this idea of hierarchical organization into the ethological mainstream, when he proposed it as one of the principles of the Grand Theory of instinct (see **3.1**). This was unfortunate because "it came to grief in the general, deserved destruction of simplistic energy models", although it was "a much more powerful principle in its own right" (R. Dawkins 1976*a*). The principle is now almost universally accepted in general biological theory and in behavioural theory specifically (Bullock 1957, 1965; Dethier & Stellar 1961; Medawar 1969*b*; Tavolga 1969; Tinbergen 1969; Hinde 1970, 1982, 1990; Anderson 1972; Ayala & Dobzhansky 1974; Fentress & Stilwell 1974; Baerends 1976; R. Dawkins 1976*a*; Bunge 1977; Granit 1977; Allen 1978, 1983; Gallistel 1980; Huntingford 1980; McFarland 1981; M. S. Dawkins 1983; Halliday & Slater 1983; Buss 1987; Szentágothai 1987; Weiskrantz 1987; Greenberg & Tobach 1989). The idea of a one-way downward flow of control as in Tinbergen's original hierarchy of behavioural control has had to give place to a much untidier hierarchy defined as web-like (Tinbergen 1969), overlapping (R. Dawkins 1976*a*) or lattice-like (Gallistel 1980) and including feedback from lower to higher levels. Several different kinds of hierarchy have been distinguished (R. Dawkins 1976*a*) but a hierarchy of integrative levels of causes is the one biologists favour.

> "Most biologists seem to agree that things, and in particular things of concern to biologists, are found not pell-mell but rather in levels... the systems at any given higher level are composed of things belonging to the immediately preceding level.... This is... the hierarchical principle... higher levels emerge out of lower ones in a natural process of self-assembly". (Bunge 1977, pp. 503–4)
>
> "Knowledge of the laws of the lower level is necessary for a full understanding of the higher level; yet the unique properties of phenomena at the higher level can not be predicted, *a priori*, from the laws of the lower level. The laws describing the unique properties of each level... express the new organizing relationships of elementary units to each other". (Novikoff 1945 in Schleidt 1981)
>
> "Behaviour has emergent properties not found in neurones, and the successive levels of interactions, inter-individual relationships and social structure have each further properties, properties which simply do not apply to the behaviour of individuals in isolation". (Hinde 1982, p. 153)

That "science must concern itself with phenomena at many levels" has been the recurrent theme of Hinde's work, referring now (Hinde 1983, 1990) to the hierarchy from molecules and cells all the way up through human societies. Referring more particularly to the mechanisms of animal behaviour, the neurophysiologist Bullock (1965) had written 25 years earlier that "nature has heaped level upon level of superimposed nervous structure". Many more clothes have since been put on that idea (see e.g. Hundert 1987) including demonstration of the hierarchical organization of cortical areas in monkey brains (Passingham 1985).

Summing up the consensus, the causal mechanisms of behaviour form an unbroken continuum of integrative levels. Each level is made up of parts found at lower levels too but nevertheless constitutes a whole that is not just the sum of its parts, because new properties emerge from the organized relations between its parts, their interaction and integration (Bunge's "self-assembly"). The next lower level, the individual parts, naturally cannot exhibit those properties. It follows that one cannot expect to discover the mechanism of some behaviour among the mechanisms operating at any much lower integrative level, jumping all the levels in between and thus ignoring the continuity and also the emergences occurring all through the hierarchy. Bullock (1958) emphasized this repeatedly from a neurophysiologist's angle, and it was very well understood by Tinbergen (1954, p. 115), who said that the tendency to "try to arrive at an understanding of the causation of behaviour by jumping to ... the level of the neurone, or of simple neurone systems ... is extremely harmful ...".

It is harmful, as M. S. Dawkins (1986) has since explained, because "circuit diagrams of the nervous system ... would be baffling in the extreme" (ibid., p. 970) and "the machinery is so complex that looking at the millions of parts that go to make it up will only confuse and make it more difficult to see how the whole works" (ibid., p. 98). "Almost never can a complex system of any kind be understood as a simple extrapolation from the properties of its elementary components" (Marr 1982 quoted by Halliday & Slater 1983). R. Dawkins (1976*b*, p. 7) coined the memorable phrase "the neurophysiologist's nirvana", arguing like Bullock that "the complete wiring diagram of the nervous system would not constitute understanding of how behaviour works.... Real understanding will only come from distillation of general principles at a higher level...". Dawkins's gibe neatly encapsulated Tinbergen's view that the wrong way to determine behaviour mechanisms is to try to work from the bottom up.

Now explicit anthropomorphists, of course, cling to the dualist idea that when one gets to the topmost, behavioural level, cognitive processes come in. This makes a real break in the hierarchy because, as the cognitivist ethological philosopher Harré (1984, p. 97) justly said, "there is no way for reconciling causal with intentional explanations within the same conceptual system". Harré, however, proceeded to argue on the lines we have already encountered from Baker (pp. 62–3) and Goodall (p. 121), offering us no choice but to accept the cognitivist, anthropomorphic hypothesis by mentioning only one alternative, namely "a descriptive vocabulary loaded with the assumption that animals are automata" (Harré ibid.). That is a reductionist caricature of neobehaviourists' views which they can be relied upon to reject.

It is not surprising that explicit anthropomorphists should think in that way, but neobehaviourists play into their hands when they too seem to envisage a break in the hierarchy at the level of whole-animal behaviour. Tinbergen must have noticed this danger for he strongly denied that there was any break:

> "The traditional distinction between 'physiology' of sense organs, nervous systems and effector organs, and the simplest types of their combined action on the one hand, and that part of ethology which is concerned with underlying causation on the other... is an artificial and outdated one, which tends to hamper research. 'Ethologists'

and 'neurophysiologists' differ, or ought to differ, with respect to the level of the observed phenomena only, and not in any other respects". (Tinbergen 1954, p. 115)

He made this point again later and this was recalled by Macdonald & M. S. Dawkins (1981): "Niko Tinbergen pioneered the study of the mechanisms of behaviour without damaging the animal in any way – physiology without breaking the skin as he aptly called it".

Nevertheless, it is still conventional among ethologists to speak of physiology as if it were something that can be studied only by breaking the skin to look at the workings of the individual parts inside the animal, not by analysing the integration of the whole animal's behaviour. Most ethologists have little taste or time for looking inside. They prefer the other approach, treating the animal as a 'black box'; "an initially mysterious object which is not to be opened but whose workings can be deduced from what it is capable of doing ... deducing what the mechanism is from the behaviour that results from it" (M. S. Dawkins 1986, p. 91). M. S. Dawkins (1983, 1986) gave some beautiful examples of the success of this black-box procedure in establishing what Staddon (1983) called "the rules of interaction" between behavioural systems. However, what does not fit with Tinbergen's conception above is that these rules are never called *physiological* rules, although the 'workings' of an animal, as opposed to its anatomy, are physiology by definition. Instead of these rules being accepted as physiological in nature, black-box studies have been described as "done without the aid of any direct physiological measurements at all" (M. S. Dawkins 1986, p. 84).

Halliday & Slater (1983, p. 4) also made an important distinction between behaviour and physiology, explaining editorially that "our authors ... are united in approaching behaviour at its own level rather than simply as a projection of physiology" (ibid., p. 4). Behaviour is indeed not simply a projection of events at lower levels of the hierarchy, but behavioural interactions nevertheless arise from physiological integration. Halliday & Slater (ibid.) said "to study behaviour at its own level is to treat the animal as a black-box, ... without being concerned too deeply with the exact nature of the intervening mechanisms" (ibid., p. 1). This seems to imply that students of behaviour have a choice when in fact they simply cannot afford to be diverted by concern with the deeper, lower-level mechanisms. The hierarchical organization requires one to start from the higher levels. The causal hierarchy of behaviour is by its very nature physiological all the way from single neurones up to behaviour. The integration of the whole animal's various behaviours, which black-box analyses display, is nothing more or less than the top level of that physiological hierarchy, as Tinbergen insisted.

This specific point is rarely spelled out, perhaps because it seems obvious after a moment's thought; but the anomaly of using a different kind of name for the top level of the causal hierarchy remains. Ethologists say that they use that name, motivation, as a label for physiological processes (Halliday 1983), yet they do not call black-box analysis physiological analysis. For example, M. S. Dawkins described Kovac & Davis's (1977, 1980) work with the sea-slug, *Pleurobranchaea* as one of the rare cases where "a motivational analysis of the behaviour of a whole animal can be followed up almost immediately by the discovery of its physiological basis" (M. S. Dawkins 1986, p. 96). "Knowing that feeding 'motivation'... inhibited 'righting' and 'withdrawal' led to the search for the physiological mechanism by which this is achieved" (ibid., p. 97). Those formulations plainly present black-box ('motivational') analysis of whole-animal behaviour as something that precedes, but is not yet physiological analysis.

Distinguishing black-box analysis by a different name, 'motivational analysis', lacks logic and makes a false separation of the whole-animal level from the lower levels of the hierarchy (a point discussed further in **4.3**).

Thus the concept of animal motivation breaks the continuity of the physiological hierarchy. The false dichotomy between physiological and so-called motivational analysis actually underrates the value of black-box analyses of behaviour as the first step in the analysis of its underlying causes. Analysis of the underlying causes must start at the top with the whole animal and proceed downwards through the physiological hierarchy level by level, since emergences often render higher levels unpredictable from lower ones. "A detailed study of the organization of behaviour at the behavioural level should precede any physiological study since this is the only way we can identify and characterize the phenomena which require explanation in hardware terms" (Huntingford 1984, p. 49). (Note that while Huntingford appreciated the need to start the causal analysis of behaviour with analysis of its organization at the top level, she still followed convention in stopping short of identifying it as physiological.) "The problem at any level of analysis is to see how the units work together to produce the emergent or resultant properties they lack singly.... If it is the relations between neural mechanisms that produce the shifting singleness of action of the whole animal, then reflex analysis of those relations provides the appropriate method. Until sufficient results are available from it the search for neural mechanisms must be blind" (Kennedy 1967, p. 261).

Reflexes are physiology, and Kovac & Davis deduced the central interactions between the different reflexes in the intact, whole sea-slug without having to break the skin or look inside the animal at all (M. S. Dawkins 1986, pp. 94–5). Looking inside came afterwards. This was equally true in Dawkins's other examples, genetics and bat echolocation (ibid., pp. 86–90) and even with "Horton the elephant" (M. S. Dawkins 1983, pp. 87–9). Looking inside merely took the analysis one level deeper into the causal hierarchy, into the neuronal apparatus of the interactions already revealed by the black-box experimental analysis of the intact animal's behaviour. What the black-box analysis and the neurophysiological analysis revealed were both physiological processes. They differed only in dealing with different levels in one and the same causal hierarchy, as Tinbergen said.

At first sight the difference between those two approaches to the causation of behaviour might appear to be trivial, merely verbal or methodological. But the way they are described shows that they differ conceptually as well. Unlike physiological concepts, the concept of motivation, the subject-matter of black-box analysis (M. S. Dawkins 1986, quoted above), is not only derived from human psychology but "was and always has been a vague term", and for that reason a multiply confusing one, too (ibid., p. 96). It may be said that it does not matter whether the results of a black-box analysis are described in motivational or physiological terms and this is true if you are interested exclusively in problems of *ultimate* causation. But the artificial break in the proximately causal hierarchy does harm by hindering the desired closing of the gap between "what nerve cells do and how animals behave". If psychological concepts had been avoided in describing the results of black-box analysis, the physiologists could have recognized these results more easily as "phenomena which require explanation in hardware terms", in Huntingford's words, and should not have found the results so difficult to use as a point of departure for carrying the causal analysis down to lower physiological levels.

In fact, black-box analyses form the logical point of departure for such downwardly stepwise, physiological probing of the hierarchy, but since the results of such analyses are not thought of as physiology they have not been used in that way. The sort of problem that could now be tackled at the next lower level might be the neurophysiological mechanism of the 'behavioural final common path'. Instead, however, of such a physiological progression there has been a proliferation of models of various types. They have been usefully surveyed by Huntingford (1984), who found them uninformative physiologically: "Even where detailed motivational models are available which accurately account for the behavioural phenomena, as in the case of the courtship of newts, these explanations... lack any information about the physiological nature of their important variables; what exactly is 'hope' in a courting male newt?" (Huntingford 1984, p. 97). In the same vein Hinde (1970) advised that "hypotheses must be judged not only at the behavioural level, but also in terms of their compatibility with lower ones. Indeed, a theoretical system which by its nature cannot be related to physiological data is unlikely to have a wide validity even at the behavioural level" (ibid., p. 7). "The student of behaviour... will be wise to aim at an analysis of behaviour into units which the physiologist will be able to handle" (ibid., p. 9). But that advice has been followed only to a limited extent: "whilst the neural bases of some of the simpler behaviour patterns of invertebrates are becoming quite well understood, we are far from comprehending the mechanisms of even such well-studied patterns as eating or sexual behaviour in higher vertebrates... some ethologists and zoologists took a rather different route, attempting to specify behavioural processes in systems theory terms.... But whilst this proved a powerful tool for increasing understanding at the behavioural level, it was less successful in paving the way for neurophysiological analyses" (Hinde 1982, p. 171). Thus the anthropomorphic break in the causal hierarchy has been costly.

While rejecting teleology and intentionality in principle (see p. 29) McFarland nevertheless makes free use of motivation and other psychological terms in describing animal behaviour (see **7.4**). He says that this is only a tactic (what is here called 'mock anthropomorphism' (pp. 88–9)), but for him physiology seems to mean an integrative level well below that of black-box analysis, as it does for cognitivists (Baker and Harré, e.g. p. 127). From those low levels one can hardly hope to unravel the mechanisms of behaviour, as we have already seen (Tinbergen 1954, p. 115; see p. 126 above, and the next paragraph). That presumably explains why McFarland despairsof most behaviour ever being explained physiologically: "we are interested in aspects of animal behaviour that are far too complex to be accounted for in physiological terms...it is doubtful that we shall ever see a complete physiological explanation of the mechanisms of anything but the most simple aspects of animal behaviour" (McFarland 1989*a*, p. 113). Huntingford (1980) takes a less reductionist view of physiology and is more optimistic: "I recognize the importance of...providing a detailed analysis and clear statement of the end products of motivational processes.... However, I look on them as stepping stones on the way to physiological models, and although I accept that these will be hard to arrive at and, perhaps, difficult to understand when we get them, I do not believe that either collecting or interpreting the necessary physiological evidence is beyond human capability".

An analytical level is the equivalent of an integrative level for research purposes. Here again there is sometimes an unnoticed break between the topmost level and lower ones. Dennett's (1987) "intentional stance' (**5.1**) is the top level of what appears at first sight to be a hierarchy of three levels of analysis of proximate causes, with analysis from a 'physical stance' at the bottom and from a 'design stance' in the middle. The meaning of the latter two stances is evident from this passage: "One could have predicted that [frog's] leap if...one had calculated the interactions from a functional blueprint of the frog's nervous system. That would be prediction from the design stance. In principle, one could have... known enough physics to predict the frog's leap from a voluminous calculation of the energetic interactions of all the parts, from the physical stance" (Dennett 1987, p. 109). The 'physical' and 'design' stances deal unequivocally with proximate causes. But the intentional stance does not do that and therefore does not take our analysis to a higher level in the same analytical hierarchy of proximate causes. In adopting the 'intentional stance' we are only pretending to deal with proximate causes, in order to gain the 'heuristic value' that this pretence yields (see pp. 88–9). For this pretence we borrow the conjectured ultimate causes of the given behaviour. In the hierarchy of real proximate causes of animal behaviour intentionality has no place.

The hierarchy of proximate causes is again in question in this passage by Hinde: "the student who aims to pursue his analysis from the behavioural level to the physiological one is exposed to a special danger – concepts useful at one stage in the analysis may be misleading at another. A classic example of this... is the concept of 'drive', 'urge', or 'tendency', which is useful at an initial behavioural level of analysis but can become a handicap at a physiological one" (Hinde 1970, p. 8, and see also ibid., pp. 199–201). The danger warning is clear there, but the grounds for it are unclear because we seem to be dealing here with a causal–analytical hierarchy of levels when in fact we are not. The 'drive' concept, which Hinde accepted for use at the top, behavioural level of that hierarchy, sounds like a proximate cause, but 'drives' are customarily identified by the function of the behaviour, its ultimate cause as Hinde himself noted (see p. 54, above). This breaks the continuity of the causal–analytical hierarchy and therefore has no place in it. Study of the function of the behaviour is useful but not because it will tell us anything whatever about the proximate causes of what the animal does. In analysing a causal hierarchy, on the other hand, the drive concept is entirely out of place because it is not a physiological concept. Andrew (1972) shared Hinde's criticism of 'drives' and 'tendencies' as causal postulates but understandably denied that they were necessary or useful even at the preliminary stage of analysis. Unfortunately that did not bring to an end the anthropomorphic "habit of giving names to systems characterized by an achievement" (Tinbergen 1963, p. 414), which is still ingrained (see **3.6**).

6.2 Displacement

Activities described as "displacement" (Tinbergen 1952) were a feature of the original Grand Theory of instinct (see **3.1**) and, as M. S. Dawkins (1986, p. 80) said, the term "should have been abandoned when the drive and energy models themselves were found to be so seriously lacking as explanations of behaviour", but instead it has lingered on. The term "has passed firmly into our language" in spite of the facts that it "declares its dependence on the Lorenz model of instinct loudly and clearly whenever it is used", and that "we should not jump to the conclusion that the behaviour we are watching is 'irrelevant' just because we cannot immediately see what the relevance is" (see examples in M. S. Dawkins 1980, p. 75). The important question is, then, why has this particular term passed into our everyday language? The evident reason is that we find 'displacement' so apt as a description of how we ourselves react when faced with a baffling situation. We are very familiar with the helpless, 'inhibited' feeling induced by such a situation and with the involuntary urge to do something quite irrelevant such as scratching the head or inspecting the fingernails. We are being unwittingly anthropomorphic, therefore, if we apply the same term 'displacement activity' to something we see an animal start doing when it appears to have been balked in some way described as either 'thwarted' ('frustrated') or faced with a 'conflict situation' in which the external stimuli for two incompatible behaviours are being presented simultaneously (see Hinde 1982, pp. 66–7; McCleery 183; McFarland 1985, pp. 38ff).

Although the term 'displacement activity' has not been abandoned entirely even among scientists, its original association with energy-flow models was abandoned in favour of disinhibition hypotheses of the mechanism (McFarland 1966*a*, 1969). This has only generated new problems. In fact, an anonymous referee was quoted by Roper & Crossland (1982) as complaining that "the whole area is a terrible mess... one's mind is buzzing with confusion". Much of the trouble still appears to arise from the use of anthropomorphic concepts in describing and explaining what happens.

The once generally accepted proposal of Andrew (1956), van Iersel and Bol (1958) and Sevenster (1961) was that when two strong and antagonist reactions are each sufficiently stimulated to inhibit the other, they mutually cancel out as it were, so that a third, weaker reaction is disinhibited and appears as a displacement activity. However, mathematical modelling of that situation has failed to generate displacement activities (Ludlow 1980), and this theory envisaging deadlocked reciprocal inhibition seems to have been merely a more physiological-sounding version of the original 'psychohydraulic', blocked-flow–overflow theory. It still carried the gratuitous and unphysiological implication that when one system is inhibiting a rival one, that rival could at the same time actively inhibit the first in return. Nor is there any physiological reason why a system that is suppressing a strong rival should therefore be less able to suppress a weaker one. The inspiration for this theory seems not to be physiology but the human social situation presented in comic form when two hefty bullies square up to each other enabling the puny Charlie Chaplin to make good his escape.

Furthermore the subjective, anthropomorphic terms 'thwarting' and 'frustration' appear to have created an unnecessary problem for McFarland, who could see "no *a priori* reason to suppose that thwarting of a single activity would lead to disinhibition" (McCleery 1983), although objectively, in this case, too, the animal is receiving conflicting stimuli for incompatible responses (say, recoiling from an obstacle encountered while approaching some 'consummatory' situation). To solve this problem McFarland (1966*b*; 1981, p. 220; 1985, p. 387) brought in another subjective, psychological concept, "*selective attention*", as an additional step in the causal process ending in the disinhibition underlying displacement. This he said was "closely related to the concept of searching image" (McFarland 1981, p. 27) which is also a subjective concept and now recognized as misleading (see **3.3**). Invoking, as well, the Mittelstaedt/von Holst theory of re-afference that is compared with an efference copy (**4.3**) he proposed that a discrepancy between these two could be the trigger for the shift of attention. His point was that such discrepancy could arise in situations of thwarting and frustration as well as situations of conflict. This was a speculative and complex theory. Behaviourally, a 'shift of attention' is something that involves inhibition of one activity and disinhibition of another, rather than being an alternative to it. Such switching is of course commonplace. "Most of the time causal factors for more than one type of behaviour are present.... Undoubtedly the commonest consequence of the simultaneous action of factors for two or more types of behaviour is the suppression of all but one of them. That behavioural inhibition of this type occurs is a matter of common observation" (Hinde 1970, p. 396). "The concept of attention... can too easily explain anything" (ibid., p. 415).

The trouble with the subjective concepts of 'thwarting' and

'frustration' is that they focus our attention on what is uppermost in our own minds in a like situation: the absence of an expected external stimulus (Kennedy 1985*b*). We tend therefore to overlook the excitatory stimuli for other reactions that are presented by the new situation and actively competing for the behavioural final common path. Responses to these new stimuli will take over that path (i.e. will be disinhibited) whenever the excitability of the previously on-going activity is reduced enough for some reason. For instance, the 'consummatory' drinking response to the stimulus of water will be highly excitable in a water-deprived rat, and so will be any associated learned responses such as approaching a water spout. So long as that system, drinking plus allied-behaviours, commands the behavioural final common path, it will keep any competitors inhibited. But if new responses are elicited and that system is now inhibited in turn, say by the rat finding the spout dry ('disappearance of the drinking stimulus'), then we could readily imagine that one or more of the competitors might be disinhibited enough to take over the final common path. This could, in principle, explain the displacement activity consistently observed by Roper (1984) in rats presented with the 'thwarting' situation of a dry drinking spout (Kennedy 1985*b*).

McFarland (1969, 1985 pp. 388–91) employed the term disinhibition "only in the weak sense that occurrence of one activity results (passively) from changes in the CFS [causal factor strength] of another activity" (Roper & Crossland 1982). He distinguished as 'competition' the occurrence of an activity because an increase in its own CFS (excitability) enables it to break through the inhibition imposed on it by a previously more excitable, on-going activity. Disinhibition defined in McFarland's purely permissive sense fails to account for the abnormally intense nature of displacement activities which was remarked by early authors (Armstrong 1950; Tinbergen 1952).

This intensification (increased frequency, amplitude or duration of actions) was quite consistent with the early blocked-flow–overflow theory of displacement activities but none of the disinhibition theories that superseded it took any account of the intensification. Roper (1984) was the first to provide quantitative confirmation of the heightened intensity of a displacement activity, in his water-deprived rats. This offers a possible way out of the confusion in this subject. A disinhibition theory with the addition of a rider to the effect that the disinhibited activity shows post-inhibitory rebound (pp. 142–3) would account for displacement activities in both 'thwarting' and 'conflict' situations.

6.3 Inhibition

This is a term borrowed from neurophysiology, and has sometimes been used also at the behavioural level redefined as follows: "Behavioural inhibition is said to occur when the causal factors otherwise adequate for the elicitation of two (or more) type of behaviour are present, and one of them is reduced in strength because of the causal factors for the other" (Hinde 1970, p. 396). That must be a very common occurrence, as Hinde noted (see p. 137, above), but students of animal behaviour have somewhat neglected it and once again this seems to be attributable to a kind of unwitting anthropomorphism.

The neglect is anything but new. It struck Sherrington (1913, p. 252): "The powers of the environment to incite through the nervous system the activity of this or that bodily organ have long been studied; less so its powers through the nervous system to check and restrain the bodily activities". That remark appeared in a major paper entitled "Reflex inhibition as a factor in the co-ordination of movements and postures", where Sherrington demonstrated *in extenso* that

> "the two processes of reflex excitation and reflex inhibition are to be regarded as coequal in their importance for coordination". (Ibid., p. 272)

Ethology is the study of coordinated movements and postures and yet Ludlow is one of the few modern ethologists who has given full weight to that crucial point of Sherrington's:

> "Any group of neurons with a positive input will fire unless an equally strong negative input is applied. Inhibitory connections seem essential in any coordination system. Indeed, Maynard (1972) after a thorough analysis of a ganglion of 30 neurons in the lobster, stated that 'excitatory synaptic activity in the ganglion appears to bring component neurons to an appropriate discharge level. It does not determine pattern. Inhibition plays a dominant role in carving out patterned output, and nearly all described chemical synapses between ganglion neurons are inhibitory. If this conclusion has general significance, as Maynard suggested and Bullock (1976) agreed, we may expect the pattern of whole animal behaviour to depend greatly on inhibitory connections." (Ludlow 1980)

Hinde's textbooks, while not as definite as Ludlow on this subject, have also been exceptionally Sherringtonian in their balanced view of excitatory and inhibitory effects in behavioural causation:

> "although we understand very little about the control of such activities, it may well be that no new principles will be required. Given that the organism is active, what it does may be determined by mutually inhibitory and facilitatory effects between the various possible activities, and the strength of these may depend in part on effects consequent upon the performance of each of them." (Hinde 1982, pp. 54–5.) "The influence of one activity on another may not be purely inhibitory – one type of

behaviour may increase the strength of another." (Ibid., p. 72)
"Either positive or negative after-effects of one activity on another can come about." (Hinde 1970, p. 618.)
"There have been few studies of the detailed course of inhibition between complex responses." (Ibid., p. 399)

Those few remarks amounted to a veritable programme for research on causal mechanisms, a programme which if pursued could probably have gone far to bridge the gap between what nerve cells do and how animals behave. But unfortunately those views of Hinde and Ludlow have not been widely shared by ethologists. In the original Grand Theory of instinct Tinbergen's (1950) hierarchical scheme of the motivational system underlying a major instinct portrayed only excitation, in the form of energy ('motivating impulses'), flowing down through alternative channels to neural centres of increasing behavioural specificity. The only hint of anything like inhibition was a passive block in each channel preventing the flow of energy into it until an appropriate external stimulus removed the block. Later, when the Grand Theory had been discredited in other respects, the word inhibition still did not even appear in the index of Marler & Hamilton's (1966) 700 page book dealing specifically with the causal mechanisms of behaviour; and this was at the time when mechanisms were still the centre of ethological interest. Indeed the connotation of 'motivation', the ethologist's collective term for behavioural causation, is today still one-sidedly excitatory; that is, concerned almost exclusively with what makes the animal positively do things. It is hardly concerned with the fact that at any one moment all except one of the things that the animal is ready to do are suppressed, or with how the pattern of excitation and inhibition shifts kaleidoscopically from moment to moment as a sequence of actions proceeds.

Black-box experimental analyses have brought out the elementary fact that there are inhibitory relations between different behavioural systems (see **6.1**). This is obvious from the effects of applying the 'adequate' external stimulus 'for' each of them, meaning a stimulus that excites the system enough for it to take over the behavioural final common path. Such an excitatory stimulus for one system simultaneously suppresses (inhibits) the others and the interesting thing, as Hinde (1970, p. 618) indicated in the third quotation above, is that this inhibitory input often has a certain after-effect, either excitatory or inhibitory. Post-inhibitory 'rebound' excitation (PIR) is a matter which has much preoccupied me experimentally because of its neglected theoretical importance (Kennedy 1966, 1990). It is an established phenomenon at the behavioural level as well as at lower analytical levels (Sherrington 1947; Bullock 1965; Selverston 1985). The textbooks of Hinde (1979, pp. 616–21) and Staddon (1983, pp. 39–43) both discuss it as one of the 'rules of interaction' between behavioural systems, but it does not get a mention in Marler & Hamilton (1966), Toates (1980), McFarland (1981, 1985), Gould (1982), Houston (1982), Halliday & Slater (1983), Huntingford (1984), Slater (1985) or Ridley (1986). Manning (1979, p. 15), Roper (1985) and Colgan (1989, p. 50) do mention it but question its significance mainly because, in Roper's words, "PIR is not observed during all... transitions from one activity to another", which is a misunderstanding. There has been no suggestion that PIR does occur at all neurophysiological or behavioural transitions. On the contrary, Bullock's (1957) classic review of neuronal interactive mechanisms, for instance, mentioned the occurrence of both post-inhibitory excitation and post-inhibitory inhibition. At the behavioural level, also, a whole gamut of alternative after-effects of the inhibition of flight during a landing on an aphid's subsequent flight have been demonstrated, extending from strongly excitatory through nil to strongly inhibitory, according to the circumstances (Kennedy 1966 *et ante*; cp. Kennedy & Ludlow 1974).

The incidence of behavioural PIR may turn out to be more rather than less common than is presently believed. For one thing, "since displacement activities are apparently a diverse group of phenomena, there is no reason to suppose that the underlying mechanisms are peculiar to them" (McCleery 1983). "Conflict is almost ubiquitous; and the difference between activities occurring as 'displacement' activities and those occurring as 'transitional' activities is only one of degree" (Hinde 1970, p. 417). The cases so far recognized as PIR-type behaviour are very diverse and they may be a small sample out of a much larger number. There is also a large and diverse group of puzzling phenomena known to animal psychologists as 'schedule-induced' or 'disjunctive' behaviour (Staddon 1977). Several authors including McFarland (1970) and Hinde (1982) have suggested that this is a kind of displacement activity. At the time when Roper (1983) reviewed the subject he decided against that suggestion on the grounds that "the disinhibition idea fails to explain how a schedule-induced activity comes to be enhanced above its baseline of occurrence". Roper was using the term 'disinhibition' there in the sense accepted at the time which did not include rebound. Now that he has himself demonstrated such rebound 'enhancement' quantitatively in the displacement activity of a rat, the hypothesis that PIR is a principle of widespread occurrence, bringing schedule-induced behaviour and displacement activity, *inter alia*, under one causal umbrella, becomes a serious proposition for research.

Looking back, we may ask why there has been a relative neglect of inhibition in the proximately causal analysis of behaviour. The neglect seems to be a kind of anthropomorphism inasmuch as the causal role of inhibition in the integration of our own behaviour is, of necessity, not something of which we are spontaneously aware (Kennedy 1958, 1967). Colloquially we are of course aware of feeling 'inhibited' about doing something, and of having various 'inhibitions' or 'hang-ups'. More to the point here, we can infer that inhibitory processes do operate in our brains from the fact that we think of only one thing at a time, since that requires the suppression (inhibition) of other things that are rival candidates for attention. But that merely tells us that such a process must exist. Its role in the interaction between behavioural systems is not something of which we are directly aware when we are paying attention to one thing or shifting our attention to another. On the other hand we are well aware of the mental activity of positively attending to something, which may be described as specific excitation as the mental level. Our observation of an animal's behaviour is similarly one-sided, again of necessity. We can see and record and think about what the animal is doing, but we cannot at the same time be equally aware of all the things that, thanks to inhibition, it is not doing. Inasmuch as this is just the kind of one-sidedness that we are used to subjectively, our unawareness of inhibition as a causal factor in the animal's behaviour can be described as anthropomorphic. To remember that inhibition has an underlying causal role no less important than that of excitation does not come naturally to us: it requires a deliberate intellectual effort.

6.4 Trail-following

We are used to feeling the force of the wind because we are earth-bound creatures. People therefore find it hard to believe at first hearing that flying animals cannot feel the wind in the same way. That again is not anthropomorphism in the usual sense of "ascribing mental experiences to animals" (p. 9). But it is anthropomorphism of a sort – assuming animals sense the world as we do – to assume that flying animals can feel the wind as we do when in fact they cannot. They cannot, because they are supported by the air and therefore carried along with it when it is moving as wind. A flying animal can feel movement of the ambient air only if it is movement relative to the animal, not movement relative to the ground which is wind. It is anthropomorphism in the same unorthodox sense to assume that a flying insect finds an odour source by following the plume of odour that is carried away from the source by the wind, in the same way that we see other earthbound creatures such as an ant or a dog following an odour trail on the ground: that is, chemotactically. This idea was expressed by Shorey (1973), the father of the modern study of insect behaviour as governed by pheromones, but he soon abandoned it as far as the oriented movement of a flier upwind along the plume towards the source is concerned (Shorey 1976). Movement upwind towards the odour source is now generally attributed to optomotor anemotaxis, wherein the insect responds to visually detected lateral drift of its flight track due to a side-wind, by a 'compensatory' turn into a more upwind orientation. However, in various (not all) moths and other flying insects the upwind progression is not straight into wind but occurs as the resultant of a rather regular succession of zig-zags (Baker 1986; Carde 1986; David 1986; David & Birch 1989). It is the turn at the end of each zig and zag that has continued to be interpreted as a response to loss of contact with the odour itself; that is, as chemotaxis. Shorey argued that these zig-zags may be oriented at least in part chemotactically, in both walkers and fliers:

> "this zigzag progression of the flying insects is remarkably similar to that... [of] animals that run along terrestrial odor trails.... Perhaps both terrestrial and aerial animals sense when they are diverging laterally from the highest average pheromone concentration near the central axis of

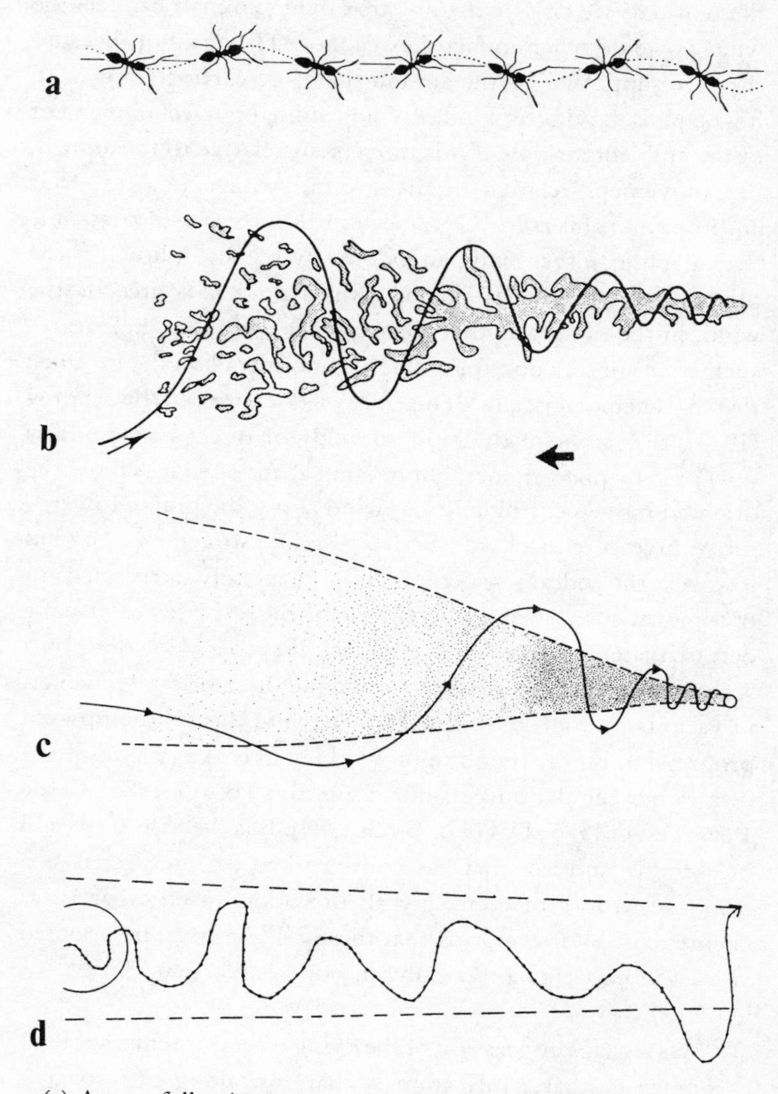

(a) An ant following a scent trail on the ground by turning back whenever it leaves the trail. (b) Flying moth following a wind-borne scent plume, hypothetically by turning back whenever it leaves the plume. Wind blowing from right to left. (c) Another the trail and are entering the lower concentrations near the boundaries ... whereupon they are caused to turn back toward the higher concentration. Such a chemotactic mechanism might continuously reorient the animal toward the trail's central axis." (Shorey, 1976, p. 24)

Shorey (1973; Farkas & Shorey 1972, 1974) had earlier adopted Butler's (1967) expression "aerial trail", remarking on the "considerable physical similarity between the elongate cloud of molecules in the air downwind from the single source, and the elongate cloud of molecules above the terrestrial trail" (Shorey 1973, p. 361). Bell & Tobin (1982) presented Shorey's ideas in more positive terms: "the mechanism for maintaining an upwind course in an odour plume has been investigated in a variety of flying insects, flying insects made to walk, and walking insects. The available evidence suggests that mechanisms of walkers and flyers are basically the same, although complicated somewhat by the third spatial dimension in flyers" (Bell & Tobin 1982, p. 246; cp. Farkas & Shorey 1974, p. 83).

In fact, the mechanisms are very different in fliers and walkers. In the first place the sensory cues used in orientation to the wind are quite different, being visual for the fliers and mechanical for the walkers. In the second place the walkers are

---

author's version of the same hypothetical behaviour. Dashed lines mark odour plume borders. (d) Actual track of a flying moth following a wind-borne scent plume (between dashed lines) and often turning back before it has left the plume. (Reproduced with permission: (a) from Schöne, H. 1984. *Spatial orientation. Spatial control of orientation in animals and men.* Princeton University Press, Princeton, NJ; (b) from Birch, M. C. & Haynes, K. F. *Insect pheromones*, Edward Arnold, London; (c) from Gerhardt, H. C. in Halliday, T. R. & Slater, P. J. B. (Eds.) *Animal behaviour*. Vol. 2 *Communication*. Blackwell Scientific Publications, Oxford; (d) from Baker, T. C., Willis, M. A. & Phelan, P. L. 1984. *Physiological Entomology* **9**, 365–76.)

continuously within millimetres of the source material (where the odour gradients are steepest and facilitate chemotaxis) along the whole length of a terrestrial trail, whereas the fliers are not in that situation along almost all of the progressively attenuating and fragmenting aerial 'trail'. Thirdly, it is not the walkers but only the fliers that are subject to wind drift and respond to it as mentioned above; and moreover they do so in a manner still not understood when they are maintaining the alternating cross-wind courses and tracks that make up zig-zagging and 'casting' (Marsh *et al.* 1978; David 1986; David & Birch 1989). The insect doing that is not, as has been thought, 'feeling for the edges' of the odour plume as we feel our way along a dark corridor by touching the walls. The oscillation between left and right turns involved in zig-zagging and casting is now known to depend on an internal counterturning generator (Kennedy 1983, 1986; Baker 1990). This is switched on by the odour stimulus and produces more frequent self-steered turns the stronger the stimulus. High-frequency self-steered counter-turning therefore produces narrow zig-zagging, and this is the only experimentally established mechanism by which a flier stays within or close to an odour plume. Likewise, widening the cross-wind casting by reducing the counterturning frequency is the flier's only known mechanism of regaining contact with a lost odour plume. Counterturns occur even within a uniform cloud of pheromone and in clean air after pheromone, where there are no gradients of odour concentration. There is no satisfactory evidence that flying insects (as distinct from walking ones: Tobin 1981; Bell & Tobin 1982) turn back when they encounter the decreasing concentration of odour as they pass outwards from the interior of a plume into the odourless air to the side. This would be a chemotactic reaction but it has not been demonstrated to play any part in a flying insect's homing-in on an upwind odour source from a distance. The small amount of experimental evidence adduced for it (Farkas & Shorey 1972; Cardé & Charlton 1985) is otherwise explicable (Baker 1990). Farkas & Shorey (1974), Birch & Haynes (1982), Gould (1982), Gerhardt (1983) and McFarland (1985) have all made the mistake of depicting the counterturns as occurring only at the edges of the plume where the flier enters clean air (p. 146). It seems safe to infer that there is an unwitting – and in a certain sense, anthropomorphic – bias in favour of chemotactic trail-following, from the way authors have figured the performance as they imagined it.

CHAPTER 7

7.1 Illusions

Most neobehaviourists (as defined on p. 6) ceased to worry about anthropomorphism years ago. They consider the battle against it was very necessary but was won by about the middle of the century. They do not believe that that way of thinking now represents any danger to the scientific study of animal behaviour. They are more concerned, especially in North America, with disposing of the last remnants of radical behaviourism. They believe that anthropomorphic language does no harm these days, at least among behavioural scientists, because they believe it is used purely metaphorically and refers only to functions, not causes. That is the view of zoological neobehaviourists at any rate, as expressed by Krebs & Davies (1981, p. 351) as we saw earlier (pp. 52–3). Also by Ridley:

> "Mechanisms and purposes [i.e. functions] have been confused particularly often. However, now the distinction has been made it causes little difficulty" (Ridley 1986, p. 7). "Subjective intentions ... are almost invariably ignored by ethologists .... It is a normal part of scientific procedure to take words from ordinary language and modify their meaning; the scientists are not confused by it, although outside commentators sometimes are." (Ibid, p. 177)

These are no doubt comfortable beliefs but they do not seem tenable in the light of the preceding chapters. I have been surprised myself to find how often scientists are unknowingly misled by the anthropomorphism of our ordinary language.

This book is short because time was limiting, not because more examples could not be found. Evidently I too have been underrating how powerful and pervasive is the influence of our 'in-built' anthropomorphism upon all of us in the field. But, after all, if anthropomorphism is built into us so that we are hardly aware of it most of the time (see **2.7**), we should not be surprised to find that we underestimate its influence. For the same reason we should not be surprised, either, to find that those who have seen through the various errors discussed in the preceding chapters seldom mention anthropomorphism, let alone identifying it as the root cause of such errors. This, perhaps more than anything, shows up how general is underestimation of the influence of anthropomorphism.

The underestimation is further highlighted if we compare Tinbergen's attitude to the perennial question of what language to use in describing animal behaviour, with the contemporary attitude as exemplified by Krebs & Davies (1987). Tinbergen 1942, 1951, 1963) was always anxious about the damaging effects of subjective teleological language on the progress of ethology: "Our habit of giving names to systems characterized by an achievement has made thinking along consistent analytical lines much more difficult than it would have been if we could have applied a more neutral terminology" (Tinbergen 1963, p. 414). But he felt unable to recommend actually using more neutral terminology to avoid the dangers of teleological language. This he said was because he found such terminology too "dry" and "non-committal", although he never seems to have suggested why this was so. Soon he turned his research attention to the functions of behaviour, a field where teleological language may be excused as mock anthropomorphism. So such language went on being used for both ultimate and proximate causes of behaviour and Tinbergen's worry about the dangers of it gradually faded among his successors. With the coming of Behavioural Ecology and Sociobiology the worry has virtually disappeared from sight – and the dangers have substantially increased.

In the Introduction to their influential book on behavioural ecology, Krebs & Davies (1987, p. 3) made a case against using "traditional formal scientific style". They quoted an eloquent passage from George Orwell to justify their decision to use a "simple direct style" and "informal shorthand" instead. However, 'simple direct' English, being good everyday English, is suffused with anthropomorphism, as we have seen (see Keeton on p. 1; Bonner, p. 25; Asquith, p. 26; Dunbar, p. 28; McFarland, p. 29). Krebs & Davies (1987) mentioned criticism of their "catchy descriptive labels" for being too anthropomorphic but dismissed this criticism in one sentence: see p. 52, above. Otherwise they mentioned only one potential disadvantage of their informal style, that "loose terminology can indicate imprecise thinking and half-formulated ideas" (Krebs & Davies 1987, p. 3). True enough; but even this disadvantage they immediately discounted by saying "it is equally easy to conceal woolly arguments behind an obfuscating screen of scientific jargon" (ibid.): which is also undeniable – and leaves us in the air. Possibly they share with a number of others authors (e.g. Kummer 1982; Gray 1987; Lockery 1989) the illusion that 'ordinary', 'everyday' or 'common-sense' language is neutral, when in fact it is suffused with anthropomorphism. That might explain why they seemed so unconcerned by the damage being done to the causal analysis of behaviour as a side-effect of the take-over by functional studies (see **3.7**). The use of subjective terminology is now widespread among neobehaviourists, and it is evident from the previous chapters that even the scientific cognoscenti cannot be relied on not to be confused by it. The stylistic issue that Krebs & Davies raised is important for its facilitation of communication, but even more important is the underlying conceptual issue that Tinbergen perceived and which badly needs to be revived.

7.2 Indulgence

The finding of substantial remnants of unconscious anthropomorphism among neobehaviourists should have been expected. More surprising and disturbing is the fact that old-style explicit anthropomorphism has actually being regaining ground in the last twenty years (p. 5, **1.1**, **4.1** and below). This calls for a fresh look at the legacy of radical behaviourism. There were two distinct sides to it. Keeton (1967; cp. Schöne 1984 and Dewsbury 1989) distinguished them as (1) oversimplification of behavioural mechanisms, and (2) opposition to anthropomorphism. Keeton identified the oversimplification as the radical behaviourists' big mistake, and the anti-anthropomorphism as their historic positive contribution to science. Having quoted Morgan's famous 'canon', "In no case may we interpret an action as the outcome of the exercise of a higher psychical faculty, if it can be interpreted as the outcome of the exercise of one which stands lower in the psychological scale", Keeton went on to say that this declaration "freed the study of animal behavior from the stranglehold of the uncritical anthropomorphic approach of earlier times and enabled it to develop as a valid branch of science. But on the debit side over-zealous application of the canon led workers to underestimate the capabilities of the complex central nervous systems in higher animals" (Keeton 1967, p. 456). Staddon (1986, p. 84) likewise faulted the radical behaviourists not for their anti-anthropomorphism but for their oversimplification in that they "emphasized the wrong aspect of reflexology, attending to the stimulus-response property of reflexes and not to reflex integration".

Nevertheless the two different sides of radical behaviourism are still being confounded. They are confounded, in particular, when it is suggested that the case against anthropomorphism has been somehow weakened by recognition of the great complexity of social behaviour among higher mammals. The philosopher Geertz (1975) saw the fear of anthropomorphism as quite gone now, exultantly dismissing it as "the dying echoes of the great mock civil war between materialism and dualism generated by the Newtonian revolution" (ibid., p. 57). Without going anywhere near that extreme Hinde also saw anthropomorphism as becoming more acceptable: "partly in accordance with the principle of parsimony, most ethologists have been unwilling to impute cognitive capacities to animals. Recent work on the complex social behaviour of higher mammals, especially primates, and even more laboratory work designed to assess the cognitive capacities of the great apes... has demonstrated that parsimony has been overdone. Fear of the dangers of anthropomorphism has caused ethologists to neglect many interesting phenomena, and it has become apparent that they could afford a little disciplined indulgence" (Hinde 1982, p. 76–8).

Hinde did not specify to what laboratory work on the cognitive capacities of the great apes he was referring there. It is likely that an important part of it was the work discussed here on ape language (**3.4**) and ape self-awareness (**5.3**). If so, that work as we saw provides no justification for indulgence towards anthropomorphism. Nevertheless a number of other authors have also called for such indulgence (e.g. Stuart 1983; Dunbar 1984*a*, p. 230; Harré 1984; Evans 1987; Tudge (quoting Bateson unpublished) 1987; Goodall 1989). The call seems most unhelpful. Who is to say how disciplined the indulgence must be? Where does disciplined indulgence end and indiscipline take over? Once we start indulging our in-built anthropomorphism it will lead us inexorably back to undisciplined, traditional anthropomorphism. That is not an imaginary danger. Burghardt had already seized upon Hinde's earlier statement that a concept like 'drive' was useful at the initial, behaviour level of analysis (Burghardt 1970, p. 8; see also **6.1**) in order to issue a call for return to the original grand theory of instinct: "Hinde (1956, 1960)...presented detailed critiques of ethological energy and drive conceptions based mainly on the limitations of the early Lorenz–Tinbergen model. However... Hinde... now is more constructive and emphasizes repeatedly throughout his recent book (Hinde 1970) that while such concepts need to be used carefully, they are useful relative to the type of analysis engaged in.... In fact, recent work indicates that the burial of the ethological motivational views was somewhat hasty" (Burghardt 1973, p. 357).

Since anthropomorphism appears to be in large measure 'human nature' (see **2.7**) our attempts to free ourselves from it are quite literally 'against human nature' and must often fail. The fear of anthropomorphism was thus a legitimate fear, even if it did cause both radical and neobehaviourists to neglect interesting phenomena. In any case, what brought the neglected phenomena to light was not the more indulgent attitude to anthropomorphism. Rather, as Hinde in fact said (see above), it was the accumulating evidence of behavioural complexity. Some ethologists have felt that this new evidence liberated them not only from the previous oversimple conceptions of behavioural mechanisms traceable to the radical behaviourists, but also from what they felt were the irksome 'killjoy' constraints of anti-anthropomorphism. If gaining the freedom to anthropomorphize about the proximate causes of animal behaviour feels to them like liberation, that testifies once again to the strength of our in-built anthropomorphism.

There is no doubt that the radical behaviourists did grossly oversimplify the mechanisms of behaviour and minimize the role of central factors. They wanted desperately to "leave no dark recesses whatever in the nervous system for demons to lurk in" (Kennedy 1958; cp. Barlow 1990). While their oversimplification was in that sense due to their anti-anthropomorphism, the reverse was not true: their anti-anthropomorphism was is no way due to their oversimplification. In short, nothing has happened to warrant more indulgence towards real (as distinct from 'mock') anthropomorphism. Something important has happened, but it is not that. Enough evidence has now accumulated to dispose once and for all of past underestimation of the complexity both of behavioural adaptations and of their causal mechanisms. That is an outcome which Tinbergen looked forward to and would have been delighted to see being achieved at last (see e.g. Tinbergen 1965).

The irony here is that the new evidence of complexity that has rendered untenable the old simplistic ideas of causal mechanisms comes in large measure from the explosion of research under the banner of behavioural ecology and sociobiology (see **3.7**). All this research on ultimate causes has now taken over an historical role once filled by explicit anthropomorphists, that of generating useful spin-off in the shape of evidence of behavioural phenomena unsuspected by the behaviourists, phenomena which now cry out for proximately causal analysis and are beginning to receive it (see e.g. Krebs & Horn 1990). Regrettably, the further irony here is that much of this new information is being diverted into the blind alley of neoanthropomorphic cognitivism.

In addition to grossly oversimplifying behaviour and minimizing the role of central nervous mechanisms, the radical behaviourists put a major obstacle in the path of research on the functions of behaviour when they triumphed over the animal mentalists, a valid point made for example by the animal cognitivists Mitchell & Thompson (1986, p. xxiv). But this again was not due to the behaviourists' anti-anthropomorphism: quite the contrary. It was hardly possible to study the functions of behaviour objectively at that time because its evident adaptiveness was commonly taken as evidence of the animal's purposiveness (see **5.1**). It had yet to be generally recognized that there were two distinct meanings of purpose: one referring to a mental, proximate cause of behaviour and the other to adaptedness resulting from natural selection. Progress in the objective study of function had to wait for the discrediting of that first, anthropomorphic meaning of purpose as intentionality on the part of the animal. It was the radical behaviourists' freeing of the subject from anthropomorphism that made possible the eventual flowering of behavioural ecology as the objective, quantitative, neobehaviourist study of the adaptive functions of behaviour (see **3.7**). This liberating role of radical behaviourism has been rather lost sight of in the great backlash against it, enabling anthropomorphism to creep back as animal cognitivism.

7.3 Neoanthropomorphism

Anthropomorphism has never been easy to resist but it has become still more difficult to resist now that it is, for the most part, no longer explicit. We have entered a new phase in the long-drawn-out human struggle for liberation from the constraints of ancient animism. Being in-built, the anthropomorphism of scientists has not been eliminated but merely driven underground, so to speak, and now works mostly through unconscious assumptions. This is what is meant by the term 'neoanthropomorphism'. In **2.7** it was said that the primary aim of this book was not to demonstrate that animals are unconscious but rather to bring out the danger of unthinkingly assuming that they are conscious. This applies equally to many of the other contested concepts. The primary aim here is not to demonstrate that animals are not self-aware, do not think, do not have purposes, do not use mental images of goals to attain them, etc., even if all that be true. More pressing is the danger of mistakes arising from assuming animals *are* conscious, *are* self-aware, *do* think, *do* have purposes, *do* use mental images, etc. without having ruled out alternative explanations of the animals' behaviour. When such assumptions are more or less unconscious one cannot expect alternative explanations will come to mind. We are doubly tempted into anthropomorphism because we for our part are predisposed to think that animals have minds like ours and they for their part seem to confirm this by "acting as if they have minds" (Gallup 1982) thanks to the optimizing effect of natural selection on behaviour.

The fundamental difficulty, however, when we wish to avoid anthropomorphism, lies in the nature of our ordinary language. When Tinbergen counterposed teleological and neutral language (see **7.1**) he was referring to the dilemma that all neobehaviourists face. On the one hand, terminology that is 'neutral' compared with our everyday language, in other words free of teleological, anthropomorphic overtones, is much to be desired for the scientific study of animal behaviour. But on the other hand such wholly objective language is almost impossible to achieve completely and attempts at it are usually clumsy and prolix because they are inevitably strained compared with our everyday speech (see **2.6** and **2.7**). Terms such as 'mind-reading' and 'manipulation' which R. Dawkins & Krebs (1978) and Krebs & R. Dawkins (1984) applied to communication between animals, are "catchy" as they said (see **3.6**). But there is a price to pay: these terms are catchy precisely because they are anthropomorphic (see also critiques of such language by Estep & Bruce 1981; Gowaty 1982; Smith 1986; McFarland 1989*c*). The terms are unexpectedly and gratifyingly familiar to us from our subjective experience – an "*Aha! Erlebnis*" (to borrow the expression used by Lorenz (1950,

p. 266), in the absence of an English equivalent), making it all too easy for us to forget that they are only metaphors. Because of that gratification, metaphorical or 'mock anthropomorphism' is sure to go on being used in discussing animal behaviour, even by people who acknowledge it is dangerous, because it is not only useful but also vivid and memorable. It will go on being used for the same reason that we use it when we are talking of inanimate systems. Then, it is pure gain with no insidious side-effects. Our everyday language would be crippled without its constant use of metaphors and analogies. Most people would not claim to have understood anything about the transmission of sound or radio signals, for example, if they had not had it explained by the analogy of waves travelling across a water surface, with which they are familiar from direct experience. Anthropomorphic analogies for animal behaviour are the exception: they readily generate misunderstanding.

All in all, the weight of our in-built anthropomorphism is formidable, acting as it does through the very language we use for thinking. Tinbergen sensed that the problem lay with our language, but this problem defeated him (p. 151; also Kennedy 1987*a*) and the eventual result has been the recent resurgence of anthropomorphism is various guises briefly documented in this book. The general unawareness of the present still-weighty influence of anthropomorphism seems to extend into Medawar's (1969*a*, *b*) widely read exposition and endorsement of Karl Popper's anti-Baconian, anti-inductivist view of scientific method (cp. Keynes 1967). One of their counts against inductivism was that it "provides no acceptable theory of the origin and prevalence of error" (Medawar 1969*b*, p. 33). But in anthropomorphism we have an obvious, major source of error in the scientific study of animal behaviour. To be so oblivious to anthropomorphism is to be unwittingly touched by it oneself. Medawar followed Popper in rejecting the "traditional inductive theory... that error must be held to arise from misapprehension of the facts ... through blindness or prejudice" (ibid, p. 32). But anthropomorphism is in fact a kind of blind prejudice, and that is why it gives rise to errors. We can, therefore, after all, offer an acceptable theory (or at any rate part of one) of the origin of inductive errors, namely, that they are, like Animism, intrusions into science from outside it. Scientific thought is not sealed off from the rest of our culture and has to deal constantly with such intrusions, like morphic resonance, 'unidentified flying objects' and Creationism.[1]

7.4 Prescription

McFarland's blunt description (*in litt.*) of anthropomorphism as an 'incurable disease' is apt enough; but it is unduly pessimistic. Symptoms of this malady do indeed appear whenever we set about analysing the causes of behaviour and we shall probably never throw it off entirely, but we already have evidence that the disease need not be fatal. Whatever its faults radical behaviourism did put a historic check on once-rampant anthropomorphism. Neobehaviourism should be able to check it further without having the counter-productive side-effects that radical behaviourism had. But that will happen only if we can do more to limit the harm that is evidently still being done

[1] Popper's and Medawar's opposition to induction thus seems rather excessive. Certainly, scientific knowledge does not "grow out of simple unbiassed statements reporting the evidence of the senses" (Medawar 1969*a*, p. 41); we cannot look at facts without preconceptions; and we cannot hope to establish final truths. Nevertheless, few would dispute that the couplet from Wordsworth which adorned the cover of *Nature* for many years: "*To the solid ground of nature trusts the Mind that builds for aye*" remains valid. My own recipe for scientific advance is to give primacy neither to induction nor to deduction but to *alternate between them*, continuously. An hypothesis calls for new facts to test it; they either confirm it, or falsify it and lead to a new hypothesis; and so on.

by anthropomorphic language to the science of animal behaviour. Several ways of doing that can be suggested. Of course, warnings are still pronounced against one or other manifestation of anthropomorphism from time to time, but these are usually voiced briefly *en passant* and without concrete recent examples of resulting mistakes. It is not enough to say simply that motivation is a concept that must be used with care. We have to acknowledge now that such warnings have not been enough to halt the tendency to retreat from unequivocal neobehaviourism towards anthropomorphism.

First, although the temptation to use everyday language is greater in the study of animal behaviour than in other branches of natural science because that everyday language is anthropomorphic, the temptation can be resisted by stricter adherence to the normal practice in other branches of science, irksome though this may be. While "everyday (non-scientific) language is riotously rich, flexible, and contextually nuance-ridden.... The scientific conceptual apparatus, by contrast, tries for economy, precision (*in*flexibility), and context neutrality. Thus, if science adopts everyday terms for its own purposes... they must then be adapted... to eliminate their context dependency, vagueness, and imprecision" (Wilkes 1988, p. 33). Every scientific discipline has its core set of special meanings of words and neologisms that cannot be dismissed as *unnecessary* jargon. The study of animal behaviour is no exception to this, but it is exceptional among the sciences in that it cannot take over everyday words with impunity. When it does, there is the penalty to pay that they are open to anthropomorphic misinterpretation, conscious or unconscious. In seeking therefore to adapt subjective terms for safer use in the science of ethology, we may not be able to free them completely from anthropomorphism which sees animal behaviour in the inappropriate context of human behaviour, but we can improve upon some of these terms if we stop to think. They are likely to have been adopted because of their (often unconscious) subjective appeal rather than because there is no adequate alternative. So, as Gowaty (1982) pointed out with regard to anthropomorphic sexual terms, we can often pick more objective terms – with some loss of vividness perhaps, but not necessarily any loss of crispness. We can very well describe an animal as 'balked' instead of 'frustrated' or 'thwarted', or as 'scanning' its environment instead of 'searching', or as making an 'incipient movement' instead of an 'intention movement'. In each case the replacement term has the advantage of directing one's attention to the actual behaviour. Admittedly, Estep & Bruce's (1981) well-intentioned attempt to avoid the anthropomorphism of 'rape' by replacing it with the descriptive term 'resisted mating' did not quite succeed, but this is merely because the new term does not cover unresisted initiation of copulatory behaviour without prior courtship.

Secondly, it is often difficult nowadays to know whether an author who uses anthropomorphic language means it literally or not (p. 90). Authors may not be clear themselves as to which they mean, not stopping to ask themselves. So when such language is being used merely as an 'intentional stance', that is, as 'mock anthropomorphism' (p. 88) with no wish to imply genuine anthropomorphism, then authors should make a habit of spelling this out. It would be a great help if they would insist, more often than they do, that the language was purely metaphorical. For instance, it is quite clear from the quotation on p. 151 that R. Dawkins regards anthropomorphic language as purely metaphorical, and he expressly distinguished "conscious intention to deceive" from "an effect functionally equivalent to deception" (R. Dawkins 1976*b*, p. 68). Yet Mitchell (1986) remarked that in spite of this R. Dawkins & Krebs (1978, p. 302) defined deceit as "deliberate misleading", which may well be read as an anthropomorphic formulation because they do not explicitly say otherwise. It would also help to put in, from time to time, a translation of what has been said into strictly objective terms no matter how clumsy and long winded. It cannot be left to the reader to take the trouble to do that unbidden. Obviously it would defeat our object if such statements appeared so often as to become merely boring and irksome to the reader. But equally obviously they do not appear often enough nowadays to prevent even sophisticated people from consciously or unconsciously taking anthropomorphic language literally.

McFarland's handling of the problem of anthropomorphic language is also ambiguous. Again, it is clear that in principle he disapproves strongly of anthropomorphism (see above and pp. 29, 72 and 98; as well as Chapters 5 and 6 of McFarland 1989*a*). In fact he anticipated the main themes of this book when he wrote that "reports of animal behaviour frequently contain implicity anthropomorphic assumptions" and that "humans... often have to be specially trained to resist the temptation to interpret the behaviour of other species in terms of their normal behaviour-recognition mechanisms" (McFarland 1981, p. 16). Yet his readers may not always realize that that is his position because he often uses anthropomorphic, subjective terms in describing animal behaviour and control system models of it without reminding the reader that he is using the terms metaphorically only (e.g. p. 77 above) – and does this even in making the case against anthropomorphism (McFarland 1989*a*, *b*). Some of his own entries in the multi-authored *Oxford companion to animal behaviour* (McFarland 1981) do embody a caution when the term in question is implicitly anthropomorphic: *Goal-Directed Behaviour* is a case in point. However, many other entries of his explain subjective terms without such a caution: *Curiosity*, *Cognition*, *Insight* and *Intention Movement* are a few of these. This widely used work of reference of his could have served as an exhibition of identified anthropomorphisms, but does not. His claim is that he is obliged to express himself in anthropomorphic, teleological language because he simply would not be understood otherwise (McFarland 1989*c*). This seems as defeatist as his view that behaviour may never yield to physiological analysis (p. 133) and it is, in effect, another tribute to the strength of the prevailing anthropomorphic bias. Using anthropomorphic language without unmistakable qualification can hardly be expected to weaken that bias.

A third way to resist neoanthropomorphism is similar to the second, but now more specifically in the context of research and one's trying to make sense of its results. It is here that pausing to translate one's racy anthropomorphic account of a behavioural sequence into a detailed description of what was actually observed can bring to light features that the racy account missed or glossed over. "Teleological terms such as 'searching' and 'appetitive'... really describe... the animal's presumed state of mind.... What it is actually doing goes without saying – or even looking.... Trying to find an objective substitute for a teleological term will always pay off in research because it forces a mental break-out and a closer look at the components of the behaviour as actually observed, components which the handy teleological term leaves unnoticed or at best unformulated in the back of the mind" (Kennedy 1986, pp. 23–4; see also Tinbergen on p. 52). For example, translating the subjective term 'searching' into the objective one 'scanning' brings us straight down to earth into the real world of behaviour. We now have to look closely at the animal's actual movements, identify what kind they are and analyse their organization. Our unthinking assumption might well be that searching referred to movements of the whole animal, but in reality scanning may consist of movements of the head only, or of receptor-bearing organs such as eyes or antennae only, or simply of a crab's maxillipeds by means of which its chemoreceptors sample the surrounding water in different directions and thus register the direction of potential prey. We have also to check our hypothesis as to what it was the animal was searching 'for'. Translated, this means that we have to identify experimentally the situation or object to which the scanning animal was especially ready to respond and in doing so cut short its scanning movements. The value of bringing the subjective terms 'thwarting' and 'frustration' down to earth in the same sense was considered on pp. 137–8. Furthermore, in putting together causal models of behaviour we need to avoid formulations which mystify physiologists, acting on the admonitions of Hinde and others that hypotheses should be, at the very least, not incompatible with physiology.

Fourthly, in the present context of creeping anthropomorphism it would help to have a fresh, uncompromised conceptual framework of neobehaviourism, and for that the best place to start would undoubtedly be with Tinbergen in the years when he was still very much concerned with the proximate causes of animal behaviour. There have been many occasions in the preceding chapters when valuable insights of his have been brought to light again after being left aside for decades. He came to accept the main criticisms of the Lorenz–Tinbergen Grand Theory of instinct (**2.1**) and warned repeatedly against what he called, prophetically, the "tenacious hold" of subjectivism. He complained of the habit among ethologists of labelling behavioural systems according to what they achieved for the animal (above), conceived as the end or purpose of the behaviour, as if this were its proximate cause when it was in fact an ultimate one (**3.6** and **4.4**). In 1963 he still felt that "teleology ... may well be a stumbling block to causal analysis in its less obvious forms.... The more complex [are] the behaviour systems we deal with, the more dangerous this can be" (Tinbergen 1963, pp. 413–14). He defined subjectivism as "replying to the question 'What causes this behaviour?' by referring to a subjective experience", and said "ethology has not yet completely succeeded in freeing itself from subjectivism in this sense ... in its subtler forms it is still very much with us'' (ibid, p. 413).

Thus Tinbergen saw clearly the dangers of anthropomorphism and its connections with subjectivism and teleology, the dangers that are still with us today. Conversely, he would have been the first to say he had left some theoretical issues unresolved. Something he appears not to have recognized fully was that the recurrent confusion between functions and causes, which worried him so much, was rooted in anthropomorphism. To that extent, he could be described as the archetype of a neobehaviourist who is the unwitting victim of in-built anthropomorphism. Because of that, I have argued elsewhere (Kennedy 1987*a*) that it is useful to look back also to certain other scientists of Tinbergen's generation or earlier who were not naturalists but were in principle neobehaviourists, notably Sherrington, Pavlov, von Holst and Weiss. The comparative psychologists Gallistel (1980) and Staddon (1983) have already demonstrated in their well-known modern textbooks the value of turning back to those pioneers, and to Sherrington in particular, in building their own present-day conceptual structures. "Sherrington's concept of the reflex", Staddon wrote, "is far from the simple, inflexible, push-button caricature sometimes encountered in introductory textbooks" (ibid., p. 30). Unfortunately that was how Tinbergen (1951) conceived the reflexes of Sherrington and Pavlov and, thanks most of all to the radical behaviourists, it is because of that crudely simplified, mechanistic guise in which reflexes are presented that they have been written off by neobehaviourists and cognitivists (Gardner 1987; Cheney & Seyfarth 1990) to this day. Had Tinbergen not been similarly prejudiced against Sherringtonian reflex analysis he might not have felt forced to accept defeat for his endeavours to rid ethology of subjective, teleological language (Kennedy 1987*a*). Ethologists have yet to follow the psychologists' lead in this matter. Among ethologists, Manning (1979, pp. 3–15) and Slater (1985, p. 17) have invoked Sherrington for the relevance of some of his work to the study of whole animal behaviour, but they made little use of it (e.g. p. 54). In fact Sherrington anticipated the ethologists in his use of the black-box approach (p. 128) to analysing mechanisms. He did not use it to analyse the interactions among the movements of whole animals but he did use it to analyse laws of interactions among the movements of whole limbs as coordinated by the spinal cord. His importance today lies in the fact that his black-box analyses hold the key to closing the gap between what nerve cells do and how animals behave. His laws of interaction between reflexes made no reference to neurones, because they were higher-level physiological laws. From there it should be possible to work on down through the hierarchy (p. 131).

In conclusion, I think we can be confident that anthropomorphism will be brought under control, even if it cannot be cured completely. Although it is probably programmed into us genetically as well as being inoculated culturally that does not mean the disease is untreatable. We human primates can defy the dictates of our genes (see **2.4**). Anthropomorphism may be showing some resurgence just now but over the last two centuries it has been retreating. This must be credited to the remarkable human social invention called science which provides a world-wide mechanism for cross-checking one another's pronouncements against the phenomenal world about us and thus for discrediting myths. Scientific progress continues, and "scientists no more enjoy discovering they are mistaken than anyone else, but built into science are mechanisms calculated to uncover mistakes and to force scientists to change their minds" (D. L. Hull 1983). If scientists, at least, finally cease to make the conscious or unconscious assumption that animals have minds, then the consequences can be expected to go beyond the boundaries of the study of animal behaviour. If the age-old mind–body problem comes to be considered as an exclusively human one, instead of indefinitely extended through the animal kingdom, then that problem too will have been brought nearer to a solution.

REFERENCES AND CITATION INDEX

Page numbers in square brackets at the ends of references indicate citations in the text.

Able, K. P. 1980. Mechanisms of orientation, navigation, and homing. Pp. 284–373 in Gauthreaux. [Pp. 61, 65]

Alcock, J. 1979. *Animal behavior: an evolutionary approach*. 2nd Edn. Sinauer Associates: Sunderland, MA. [Pp. 39, 49, 50]

Allen, G. E. 1978. *Life science in the twentieth century*. 2nd Edn. Cambridge University Press: Cambridge. [P. 124]

1983. The several faces of Darwinism: materialism in nineteenth and twentieth century evolutionary theory. Pp. 81–102 in Bendall. [P. 124]

Amsel, A. 1989. *Behaviorism, neobehaviorism and cognitivism in learning theory: historical and contemporary perspectives*. Lawrence Erlbaum: Hillsdale, NJ. [Pp. 2, 104]

Anderson, J. R. 1984. Monkeys with mirrors: some questions for primate psychology. *International Journal of Primatology* **5**, 81–97. [P. 106]

Anderson, P. W. 1972. More in different. *Science* **177**, 393–6. [Pp. 105, 124]

Andrew, R. J. 1956. Some remarks on behaviour in conflict situations, with special reference to *Emberiza* spp. *British Journal of Animal Behaviour* **4**, 85–91. [P. 136]

1972. The information potentially available in mammal displays. In Hinde (Ed.). [P. 136]

Anon. 1991. Guidelines for the use of animals in research. *Animal Behaviour* **41**, 183–6. [P. 122]

Armstrong, E. A. 1950. The nature and function of displacement activities. *Symposia of the Society for Experimental Biology* **4**, 361–84. [P. 138]

Asquith, P. J. 1984. The inevitability and utility of anthropomorphism in description of primate behaviour. Pp. 138–76 in Harré & Reynolds (Eds.). [Pp. 7, 26, 27, 28, 70, 85, 89]

Axelrod, R. 1984. *The evolution of cooperation*. Basic Books: New York. [P. 22]

Ayala, F. J. & Dobzhansky, T. 1974. *Studies in the philosophy of biology. Reduction and related problems*. University of California Press: Berkeley, CA. [P. 124]

Baerends, G. P. 1976. The functional organization of behaviour. *Animal Behaviour* **24**, 726–38. [P. 124]

Baker, R. R. 1978. *The evolutionary ecology of animal migration*. Hodder & Stoughton: London. [Pp. 60, 61, 64]

1982. *Migration paths through time and space*. Hodder & Stoughton: London. [Pp. 57, 60, 61, 62, 63, 64, 65, 66]

Baker, T. C. 1986. Pheromone-modulated movements of flying moths, Pp. 39–48 in Payne, Birch & Kennedy (Eds.). [P. 145]

1990. Upwind flight and casting flight: complementary phasic and tonic systems used for location of sex pheromone sources by male moths. Pp. 18–25 in Døving, K. D. (Ed.) *Proceedings of the tenth international symposium on olfaction and taste*. GCS A/S: Oslo. [Pp. 103, 148, 149]

Baker, T. C., Willis, M. A. & Phelan, P. L. 1984. Optomotor anemotaxis polarizes self-steered zigzagging in flying moths. *Physiological Entomology* **9**, 365–76. [P. 147]

Barlow, G. W. 1989. Has sociobiology killed ethology or revitalized it? *Perspectives in Ethology* **8**, 1–45. [Pp. 7, 33, 56]

Barlow, H. 1987. The biological role of consciousness. Pp. 362–74 in Blakemore & Greenfield (Eds.). [Pp. 12, 20]

1990. The mechanical mind. *Annual Review of Neuroscience* **13**, 15–24. [Pp. 2, 3, 20, 156]

Barnard, C. J. 1983. *Animal behaviour: ecology and evolution*. Croom Helm: London & Canberra. [Pp. 39, 61, 82]

Bateson, P. P. G. 1968. Ethological methods of observing behavior. Pp. 389–99 in Weiskrantz, L. (Ed.) *Analysis of behavioral change*. Harper & Row: New York. [P. 33]

1990. Choice, preference and selection. Pp. 149–56 in Bekoff & Jamieson (Eds.) 1990*a*. [Pp. 89, 90]

1991. Assessment of pain in animals. *Animal Behaviour* **42**, 827–829. [P. 117]

Bateson, P. P. G. & Hinde, R. A. (Eds.) 1976. *Growing points in ethology*. Cambridge University Press: Cambridge.

Bateson, P. P. G. & Klopfer, P. H. 1989. Whither ethology? *Perspectives in Ethology* **8**, v–viii. [Pp. 33, 55, 57]

(Eds.) 1991. *Human understanding and animal awareness*. (Perspectives in Ethology, 9.) Plenum Press: New York & London. [P. 57]

Bekoff, M. & Jamieson, D. (Eds.) 1990a. *Interpretation and explanation in the study of animal behavior*. Vol. 1. *Interpretation, intentionality, and communication*. Westview Press: Boulder, CO. [P. 96]

1990*b*. Cognitive ethology and applied philosophy: the significance of an evolutionary biology of mind. *Trends in Ecology and Evolution* **5**, 156–9. [Pp. 10, 96, 101]

1991. Reflective ethology, applied philosophy, and the moral status of animals. Pp. 1–47 in Bateson & Klopfer (Eds.). [Pp. 97, 98, 101]

Bell, W. J. & Tobin, T. R. 1982. Chemo-orientation. *Biological Reviews* **57**, 219–60. [Pp. 147, 148]

Bendall, D. S. (Ed.) 1983. *Evolution from molecules to men*. Cambridge University Press, Cambridge.

Bierens de Haan, J. A. 1937. Uber den Begriff des Instinktes in der Tierpsychologie. *Folia Biotheoretica* **2**, 1–16. [P. 4]

1947. *Animal psychology: its nature and its problems*. Burrow's Press: London. [Pp. 4, 27]

Birch, M. C. & Haynes, K. F. 1982. *Insect pheromones*. Edward Arnold: London. [Pp. 147, 149]

Blakemore, C. & Greenfield, S. (Eds.) 1987. *Mindwaves. Thoughts on intelligence, identity and consciousness*. Basil Blackwell: Oxford. [Pp. 19, 20]

Boakes, R. 1984. *From Darwin to behaviourism: Psychology and the minds of animals*. Cambridge University Press: Cambridge. [P. 86]

Bolles, R. C. 1975. *The theory of motivation*. Harper & Row: New York & London. [P. 3]

Bond, A. B. 1983. Visual search and selection of natural stimuli in the pigeon: the attention threshold hypothesis. *Journal of Experimental Psychology: Animal Behavior Processes* **9**, 292–306. [P. 39]

Bonner, J. T. 1980. *The evolution of culture*. Princeton University Press: Princeton, NJ. [Pp. 7, 16–17, 25, 46, 48, 61, 95, 96]

Bowsher, D. 1981. Pain sensations and pain reactions. Pp. 22–8 in Wood-Gush, Dawkins & Ewbank (Eds.). [Pp. 117, 118]

Brady, J., Gibson, G. & Packer, M. J. 1989. Odour movement, wind direction, and the problem of host-finding by tsetse flies. *Physiological Entomology* **14**, 369–80. [P. 103]

Branch, M. N. 1982. Misrepresenting behaviorism. *Behavioral and Brain Sciences* **5**, 372–3. [P. 2]

Bullock, T. H. 1957. Neuronal integrative mechanisms. Pp. 1–20 in Scheer, B. T. (Ed.) *Recent advances in invertebrate physiology*. University of Oregon Publications: Eugene, OR. [Pp. 124, 142]

1958. Evolution of neurophysiological mechanisms. Pp. 165–77 in Rowe, A. & Simpson, G. G. (Eds.) *Behavior and evolution*. Yale University Press: New Haven, CN. [P. 126]

1965. Mechanisms of integration. Pp. 257–351 in Bullock, T. H. & Horridge, G. A. (Eds.) *Structure and function in the nervous systems of invertebrates*. Vol. I. W. H. Freeman & Co.: San Francisco & London. [Pp. 25, 57, 58, 124, 126, 142]

1982. Afterthoughts on animal minds. Pp. 407–13 in Griffin (Ed.). [P. 121]

Bunge, M. 1977. Emergence and the mind. *Neuroscience* **2**, 501–9. [Pp. 7, 27, 124, 125]

Burghardt, G. M. 1973. Instinct and innate behaviors: towards an ethological psychology. Pp. 322–91 in Nevin, J. A. (Ed.) *The study of behavior: Learning, motivation, emotion and instinct*. Scott, Foresman & Co.: Glenview, IL & Brighton, Sussex. [P. 155]

1985. Animal awareness. Current perceptions and historical perspectives. *American Psychologist* **40**, 905–19. [Pp. 24, 99, 100, 104]

Buss, L. W. 1987. *The evolution of individuality*. Princeton University Press: Princeton, NJ. [P. 124]

Butler, C. G. 1967. Insect pheromones. *Biological Reviews* **42**, 42–87. [P. 147]

Byrne, R. & Whiten, A. 1988. *Machiavellian intelligence. Social expertise and the evolution of intellect in monkeys, apes, and humans*. Clarendon Press: Oxford. [Pp. 20, 22]

Camhi, J. M. 1984. *Neuroethology. Nerve cells and the natural behavior of animals*. Sinauer Associates: Sunderland, MA. [P. 58]

Cardé, R. T. 1986. Epilogue: Behavioural mechanisms. Pp. 175–86 in Payne, Birch & Kennedy (Eds.). [P. 145]

Cardé, R. T. & Charlton, R. E. 1985. Olfactory sexual communication in Lepidoptera: strategy, sensitivity, and selectivity. *Symposia of the Royal Entomological Society of London* **12**, 241–64. [P. 149]

Carthy, J. D. 1951. Instinct. *New Biology* **10**, 95–105. [P. 35]

Cecchini, A. B. P., Melbin, J. & Noordergraaf, A. 1981. Set-point: is it a distinct structural entity in biological control? *Journal of Theoretical Biology* **93**, 387–394. [P. 74]

Chance, M. R. A. & Jolly, A. 1970. *Social groups of monkeys, apes, and men*. Jonathan Cape: London. [P. 20]

Cheney, D. L. & Seyfarth, R. M. 1985. Social and non-social knowledge in vervet monkeys. *Philosophical Transactions of the Royal Society of London, Ser. B* **308**, 187–201. [Pp. 20, 88]

1990. *How monkeys see the world: inside the mind of another species*. University of Chicago Press: Chicago. [Pp. 5, 91, 92, 97, 101, 166]

Cheney, D. L., Seyfarth, R. M., Smuts, B. B. & Wrangham, R. W. 1987. Future of primate research. Pp. 491–6 in Smuts, B. B., Cheney, D. L., Seyfarth, R. M., Wrangham, R. W. & Struhsaker, T. T. (Eds.) *Primate societies*. Chicago University Press: Chicago & London. [P. 98]

Colgan, P. 1989. *Animal motivation*. Chapman & Hall, London. [Pp. 34, 50, 55, 97, 98, 102, 142]

Collier, G. H. 1980. An ecological analysis of motivation. Pp. 125–51 in Toates & Halliday (Eds.). [P. 51]

Crook, J. H. 1960. Nest form and construction in certain West African weaver birds. *Ibis* **102**, 1–25. [P. 36]

1964. Field experiments on the nest construction and repair behaviour of certain weaverbirds. *Proceedings of the Zoological Society of London* **142**, 217–55. [P. 36]

1980. *The evolution of human consciousness*. Clarendon Press: Oxford. [Pp. 20, 105]

1987. The nature of conscious awareness. Pp. 383–402 in Blakemore & Greenfield (Eds.) [Pp. 20, 23]

Curio, E. 1976. *The ethology of predation*. Springer-Verlag: Berlin. [Pp. 38, 39]

David, C. T. 1986. Mechanisms of directional flight in wind. Pp. 49–57 in Payne, Birch & Kennedy (Eds.). [Pp. 145, 148]

David, C. T. & Birch, M. C. 1989. Pheromones and insect behaviour. Pp. 17–35 in Jutsum, A. R. & Gordon, R. F. S. (Eds.) *Pheromones in plant protection*. John Wiley & Sons: Chichester. [Pp. 145, 148]

Davies, C. 1981. Migration. Pp. 380–7 in McFarland (Ed.) 1981*a*. [P. 61]

Davis, J. M. 1981. Imitation. Pp. 298–303 in McFarland (Ed.) 1981*a*. [Pp. 46, 48]

Dawkins, M. S. 1971*a*. Perceptual changes in chicks: another look at the "search image" concept. *Animal Behaviour* **19**, 566–74. [Pp. 38, 39]

1971*b*. Shifts of "attention" in chicks during feeding. *Animal Behaviour* **19**, 575–82. [Pp. 38, 39]

1980. *Animal suffering. The science of animal welfare*. Chapman & Hall: London. [Pp. 2, 105, 114, 116, 117, 135]

1981. Searching image. Pp. 495–7 in McFarland (Ed.). [P. 39]

1982. Evolutionary ethology of thinking. Pp. 355–73 in Griffin (Ed.). [P. 118]

1983. The organisation of motor patterns. Pp. 75–99 in Halliday & Slater (Eds.). [Pp. 36, 124, 128]

1986. *Unravelling animal behaviour*. Longman Group: Harlow, Essex. [Pp. 24, 34, 39, 58, 63, 83, 126, 128, 129, 130, 131, 135]

1989. The future of ethology: how many legs are we standing on? *Perspectives in Ethology* **8**, 47–54. [Pp. 55, 56, 57, 58, 59]

1990. From an animal's point of view: motivation, fitness, and animal welfare. *Behavioral and Brain Sciences* **13**, 1–9, 49–54. [Pp. 114, 115, 119, 120]

Dawkins, R. 1976*a*. Hierarchical organisation: a candidate principle for ethology. Pp. 7–54 in Bateson & Hinde (Eds.). [Pp. 33, 124, 125]

1976*b*. *The selfish gene*. Oxford University Press: Oxford. [Pp. 14, 15, 17–18, 21, 24, 25, 52, 70, 71, 126, 162]

1989. Darwinsism and human purpose. Pp. 137–43 in Durant (Ed.). [Pp. 30, 31]

Dawkins, R. & Krebs, J. R. 1978. Animal signals: information or manipulation? Pp. 282–309 in Krebs, J. R. & Davies, N. B. (Eds.) *Behavioural ecology: an evolutionary approach*. 2nd Edn. Blackwell Scientific Publications: Oxford. [Pp. 56, 158, 162]

1979. Arms races between and within species. *Proceedings of the Royal Society of London, Ser. B*, **205**, 489–511. [P. 22]

Daykin, P. N., Kellogg, F. E. & Wright, R. H. 1965. Host-finding and repulsion of *Aëdes aegypti. Canadian Entomologist* **97**, 239–63. [P. 103]

Dennett, D. C. 1978. Beliefs about beliefs. *Behavioral and Brain Sciences* **1**, 568–70 [P. 92]

1987. *The intentional stance*. MIT Press: Cambridge, MA. [Pp. 17, 21, 66, 92, 95, 133, 134]

Dethier, V. G. 1964. Microscopic brains. *Science* **143**, 1138–45. [P. 26]

1966. Insects and the concept of motivation. *Nebraska Symposium on Motivation* 105–136. [P. 83]

1982. The selfish nervous system. Pp. 445–55 in Morrison, A. & Stick, P. (Eds.) *Changing concepts of the nervous system*. Academic Press: New York. [P. 83]

Dethier, V. G. & Stellar, E. 1961. *Animal behavior. Its evolutionary and neurological basis.* Prentice-Hall: Englewood Cliffs, NJ. [Pp. 82, 124]

Dewsbury, D. A. 1989. A brief history of the study of animal behavior in North America. *Perspectives in Ethology* **8**, 85–122. [Pp. 153]

Diamond, D. 1981. Experimenting on animals: a problem in ethics. Pp. 337–62 in Sperlinger (Ed.). [P. 123]

Dickinson, A. 1985. Actions and habits: the development of behavioural autonomy. *Philosophical Transactions of the Royal Society of London, Ser. B* **308**, 67–78. [P. 96]

Dingle, H. 1980. Ecology and evolution of migration. Pp. 1–101 in Gauthreaux (Ed.). [P. 61]

Driscoll, J. W. & Bateson, P. 1988. Animals in behavioural research. *Animal Behaviour* **36**, 1569–74. [P. 121]

Dunbar, R. I. M. 1984*a*. *Reproductive decisions. An economic analysis of Gelada baboon social structure*. Princeton University Press: Princeton NJ. [Pp. 24, 28, 87, 88, 154]

1984*b*. The awareness of animals. (Review of Griffin 1984.) *New Scientist* **104**, 37. [P. 12]

1984*c*. Learning the language of primates. (Review of Harré & Reynolds 1984.) *New Scientist* **104**, 45. [Pp. 5, 28]

1985. How to listen to the animals. *New Scientist* **106**, 36–9. [Pp. 22, 26, 27]

1989. Review of Byrne & Whiten 1988. *Animal Behaviour* **37**, 699–700. [Pp. 20, 98]

Durant, J. R. (Ed.) 1989. *Human origins*. Clarendon Press: Oxford.

Estep, D. Q. & Bruce, K. E. M. 1981. The concept of rape in non-humans: a critique. *Animal Behaviour* **29**, 1272–73. [Pp. 53, 158, 162]

Evans, P. 1987. *Ourselves and other animals*. Century Hutchinson Ltd: London. [P. 154]

Evarts, E. V. 1971. Central control of movement. V. Feedback and corollary discharge: a merging of concepts. *Neuroscience Research Program Bulletin* **9** 86–112. [Pp. 76, 77–8, 80]

Ewert, J. P. 1980. *Neuroethology*. Springer-Verlag: Berlin. [P. 58]

Farkas, S. R. & Shorey, H. H. 1972. Chemical trail-following by flying insects: a mechanism for orientation to a distant odour source. *Science* **178**, 67–68. [Pp. 147, 149]

1974. Mechanism of orientation to a distant pheromone source. Pp. 81–95 in Birch, M. C. (Ed.) *Pheromones*. North-Holland Publishing Co.: Amsterdam & London. [Pp. 147, 149]

Farrell, B. A. 1978. Some considerations in the philosophy of mind. *Behavioral and Brain Sciences* **1**, 571–2. [P. 108]

Fentress, F. C. & Stilwell, F. P. 1974. Grammar of a movement sequence in inbred mice. *Nature, London*, **24**, 52–3. [P. 124]

Fisher, J. A. 1991. Disambiguating anthropomorphism: an interdisciplinary review. Pp. 49–85 in Bateson & Klopfer (Eds.). [P. 95]

Fisher, J. & Hinde, R. A. 1949. The opening of milk bottles by birds. *British Birds* **42**, 347–57. [P. 47]

Fitzgerald, M. 1990. c-Fos and the changing face of pain. *Trends in the Neurosciences* **13**, 339–40. [P. 117]

Fox, M. A. 1986. *The case for animal experimentation. An evolutionary and ethical perspective.* University of California Press: Berkeley, Los Angeles & London. [Pp. 18, 123]

Fox, M. W. 1984. *The whistling hunters. Field studies of the Asisatc wild dog* (Cuon alpinus). State University of New York Press: Albany, NY. [P. 95]

Fraenkel, G. S. & Gunn, D. L. 1940. *The orientation of animals. Kineses, taxes and compass reactions.* Clarendon Press: Oxford. [P. 71]

Galef, B. G. 1990. Tradition in animals: field observations and laboratory analyses. Pp. 74–97 in Bekoff & Jamieson (Eds.) 1990*a*. [P. 48]

Gallistel, C. R. 1980. *The organization of action: a new synthesis.* Lawrence Erlbaum: Hillsdale, New Jersey. [Pp. 82, 124, 125, 166]

Gallup, G. G., Jr 1970. Chimpanzees: self-recognition. *Science* **167**, 86–7. [P. 105]

1971. Minds and mirrors. *New Society* **18**, 975–7. [P. 105]

1977. Self-recognition in primates. A comparative approach to the bidirectional properties of consciousness. *American Psychologist* **32**, 329–38. [Pp. 105, 106]

1982. Self-awareness and the emergence of mind in primates. *American Journal of Primatology* **2**, 237–48. [Pp. 8, 25, 87, 107, 118, 158]

1983. Toward a comparative psychology of mind. Pp. 473–510 in Mellgren (Ed.). [P. 107]

1987. Self-awareness. Pp. 3–16 in Mitchell, G. & Erwin, J. (Eds.) *Comparative primate biology*, Vol. 2. *Behavior and ecology*. Alan R. Liss Inc.: New York. [Pp. 107, 109, 110]

Gardner, H. 1987. *The mind's new science: a history of the cognitivist revolution.* 2nd Edn. Basic Books: New York. [P. 166]

Gardner, R. A. & Gardner, B. T. 1978. Comparative psychology and language acquisition. *Annals of the New York Academy of Sciences* **309**, 37–76. [Pp. 41, 44]

Gauthreaux, S. A. (ed.) 1980. *Animal migration, orientation, and navigation.* Academic Press: New York. [P. 63]

Geertz, C. 1975. The growth of culture and the evolution of mind. Pp. 55–83 in Geertz, C. (Ed.) *The interpretation of cultures: selected essays.* Hutchinson: London. [Pp. 16, 154]

Gendron, R. P. 1986. Searching for cryptic prey: evidence for optimal search rates and the formation of search images in quail. *Animal Behaviour* **34**, 898–912. [P. 39]

Gendron, R. P. & Staddon, J. E. R. 1983. Searching for cryptic prey: the effect of search rate. *American Naturalist* **121**, 172–86. [P. 39]

Gerhardt, H. C. 1983. Communication and environment. Pp. 82–113 in Halliday, T. R. & Slater, P. J. B. (Eds.) *Animal Behaviour*. Vol. 2. *Communication*. Blackwell Scientific Publications: Oxford. [Pp. 147, 149]

Goodall, J. 1986. *The chimpanzees of Gombe. Patterns of behaviour.* The Belknap Press of Harvard University Press: Cambridge, MA. & London. [Pp. 27, 105]

1989. Foreword. Pp. vii–ix in Rollin. [Pp. 120, 154]

Gould, J. L. 1982. *Ethology. The mechanisms and evolution of behavior*. W. W. Norton & Co.: New York & London. [Pp. 13, 17, 27, 105, 142, 149]
1986. The locale map of honey bees: do insects have cognitive maps? *Science*, **232**, 861–3. [P. 101]
Gowaty, P. A. 1982. Sexual terms in sociobiology: emotionally evocative and, surprisingly, jargon. *Animal Behaviour* **30**, 630–1. [Pp. 158, 162]
Granit, R. 1977. *The purposive brain*. MIT Press: Cambridge, MA & London. [P. 124]
Gray, J. 1987. The mind-brain identity theory as a scientific hypothesis: a second look. Pp. 461–83 in Blakemore & Greenfield (Eds.). [Pp. 2, 152]
Greenberg, G. & Tobach, E. (Eds.) 1989. *Evolution of social behaviour and integrative levels*. Lawrence Erlbaum: Hove, Sussex. [P. 125]
Greenfield, P. M. & Savage-Rumbaugh, E. S. 1990. Grammatical combination in *Pan paniscus*: processes of learning and invention in the evolution and development of language. Pp. 540–78 in Parker & Gibson (Eds.). [Pp. 45–6]
Griffin, D. R. 1976. *The question of animal awareness. Evolutionary continuity of mental experience*. Rockefeller University Press: New York. [Pp. 10, 12]
1978. Prospects for a cognitive ethology. *Behavioral and Brain Sciences* **1**, 527–38. [Pp. 10, 107]
1981. *The question of animal awareness. Evolutionary continuity of mental experience*. 2nd edn, revised and enlarged. Rockefeller University Press: New York. [Pp. 10, 12]
(Ed.) 1982. *Animal mind–human mind*. Springer-Verlag: Berlin.
1984. *Animal thinking*. Harvard University Press: Cambridge, MA & London. [Pp. 10, 11, 13, 14, 101, 105]
Groves, C. P. 1978. What does it mean to be conscious? *Behavioral and Brain Sciences* **1**, 575–6. [P. 105]
Guilford, T. & Dawkins, M. 1987. Search images not proven: a reappraisal of recent evidence. *Animal Behaviour* **35**, 1838–45. [Pp. 39]
1989*a*. Search image versus search rate: a reply to Lawrence. *Animal Behaviour* **37**, 160–1. [P. 39]
1989*b*. Search image versus search rate: two different ways to enhance prey capture. *Animal Behaviour* **37**, 163–5. [P. 39]
Guthrie, D. M. 1980. *Neuroethology. An introduction*. Blackwell Scientific Publications: Oxford. [P. 58]
Halliday, T. 1983. Motivation. Pp. 100–33 in Halliday & Slater (Eds.). [Pp. 81, 129]
Halliday, T. & Slater, P. J. B. (Eds.) 1983. *Animal behaviour*. Vol. 1. *Causes and effects*. Blackwell Scientific Publications: Oxford [Pp. 58, 124, 126, 128, 129, 142, 147]
1983. Introduction. Pp. 1–9 in Halliday & Slater (Eds.). [P. 81]
Harré, R. 1984. Vocabularies and theories. Pp. 90–106 in Harré & Reynolds (Eds.). [Pp. 127, 154]
Harré, R. & Lamb, R. (Eds.) 1986. *The dictionary of ethology and animal learning*. Basil Blackwell Ltd: Oxford. [P. 77]
Harré, R. & Reynolds, V. (Eds.) 1984. *The meaning of primate signals*. Cambridge University Press: Cambridge. [P. 26]
Hediger, H. 1980. Do you speak Yerkish? the newest colloquial language with chimpanzees Pp. 441–7 in Sebeok & Umiker-Sebeok (Eds.). [Pp. 44, 45]
Henry, G. 1975. Teleological explanation. Pp. 105–12 in Korner, S. (Ed.) *Explanation*. Basil Blackwell Ltd: Oxford. [P. 25]

Hill, J. H. 1980. Apes and language. Pp. 331–51 in Sebeok & Umiker-Sebeok (Eds.). (Reprinted from *Annual Review of Anthropology* (1978) **7**, 89–112. [P. 41]

Hinde, R. A. 1959*a*. Some recent trends in ethology. Pp. 461–610 in Koch, S. (Ed.). *Psychology, a study of a science: Study I*, Vols. 1 & 2. McGraw-Hill: New York. [P. 34]

1959*b*. Unitary drives. *Animal Behaviour* **7**, 130–41. [P. 34]

1960. Energy models of motivation. *Symposia of the Society for Experimental Biology* **14**, 199–213. [P. 34]

1966. *Animal behaviour. A synthesis of ethology and comparative psychology*. McGraw-Hill: New York. [P. 6]

1970. *Animal behaviour. A synthesis of ethology and comparative psychology*. 2nd Edn. McGraw-Hill: New York. [Pp. 37, 54, 59, 71, 76, 124, 132, 134, 137, 139, 141, 142, 143]

(Ed.) 1972. *Non-verbal communication*. Cambridge University Press: Cambridge.

1982. *Ethology. Its nature and relations with other sciences*. Fontana Paperbacks: Glasgow. [Pp. 14, 50, 54, 58, 71, 124, 125, 132, 136, 143, 154]

(Ed.) 1983. *Primate social relationships*. Blackwell Scientific Publications: Oxford. [P. 125]

1990. The interdependence of the behavioural sciences. *Transactions of the Royal Society of London, Ser. B* **329**, 217–27. [P. 124]

Hinde, R. A. & Fisher, J. 1951. Further observations on the opening of milk bottles by birds. *British Birds* **44**, 392–6. [P. 47]

1972. Some comments on the re-publication of two papers on the opening of milk bottles by birds. Pp. 377–8 in Klopfer & Hailman (Eds.). [P. 48]

Hinde, R. A. & Stevenson, J. G. 1970. Goals and response control. Pp. 216–73 in Aronson, L. R., Tobach, E. Rosenblatt, J. S. & Lehrman, D. S. (Eds.) *Development and evolution of behavior*. W. H. Freeman & Co.: New York. [P. 37]

Holst, E. von 1954. Relations between the central nervous system and the peripheral organs. Pp. 75, 76, 77. *British Journal of Animal Behaviour* **2**, 89–94. [Pp. 75, 76, 77]

1973. *The behavioural physiology of animals and men*. Methuen: London. [P. 75]

Holst, E. von & Mittelstaedt, H. 1950. Das Reafferenzprinzip (Wechselwirkungen zwischen Zentralnervensystem und Peripherie). *Naturwissenschaften* **37**, 464–7. [Pp. 75, 80]

Horridge, G. A. 1977. Mechanistic teleology and explanation in neuroethology: understanding the origins of behavior. Pp. 423–8 in Hoyle, G. (Ed.) *Identified neurons and behavior of arthropods*. Plenum Press: New York & London. [P. 7]

Houston, A. I. 1982. Transitions and time sharing. *Animal Behaviour* **30**, 615–25. [P. 42]

Hoyle, G. 1965. Neural control of skeletal muscle. Pp. 402–49 in Rockstein, M. (Ed.) *The physiology of insecta*. 2. Academic Press: New York & London. [P. 58]

Hull, C. L. 1943. *Principles of behavior. An introduction to behavior theory*. Appleton-Century Co.: New York. [P. 32]

Hull, D. L. 1983. Review of Kitcher, P. 1982. Abusing science: the case against creationism. *Quarterly Review of Biology* **58**, 392–5. [P. 167]

Hulse, S. H., Fowler, H. & Honig, W. K. 1978. *Cognitive processes in animal behavior*. Lawrence Erlbaum: Hillsdale, NJ. [P. 96]

Humphrey, N. K. 1976. The social function of intellect. Pp. 303–17 in Bateson & Hinde (Eds.). [P. 20]

1977. Review of Griffin 1976. *Animal Behaviour* **25**, 521–2. [P. 12]

1983. *Consciousness regained: chapters in the development of mind.* Oxford University Press: Oxford. [P. 20]

1986. *The inner eye.* Faber & Faber: London. [Pp. 11, 13, 19, 20, 21, 23, 105]

1987. The inner eye of consciousness. Pp. 377–81 in Blakemore & Greenfield (Eds.). [Pp. 19, 20, 21, 23]

Hundert, E. 1987. Can neuroscience contribute to philosophy? Pp. 407–29 in Blakemore & Greenfield (Eds.). [P. 126]

Huntingford, F. A. 1980. Analysis of the motivational processes underlying aggression in animals. Pp. 341–56 in Toates & Halliday (Eds.). [Pp. 124, 133]

1984. *The study of animal behaviour.* Chapman & Hall: London & New York. [Pp. 10, 61, 71–2, 130, 132]

Iersel, J. J. A. van & Bol, A. C. A. 1958. Preening in two tern species: a study in displacement activities. *Behaviour* **13**, 1–88. [P. 136]

Immelmann, K. & Beer, C. 1989. *A dictionary of ethology.* Harvard University Press: Cambridge, MA & London. [P. 81]

Ingold, T. 1986. *The appropriation of nature. Essays on human ecology and social relations.* Manchester University Press: Manchester. [Pp. 16, 18]

Isaac, G. L. 1983. Aspects of human evolution. Pp. 509–43 in Bendall. [P. 18]

Itani, J. & Nishimura, A. 1973. The study of infrahuman culture in Japan. A review. Pp. 26–50 in Menzel, E. (Ed.) *Precultural primate behavior.* S. Karger AG: Basel. [Pp. 46, 48]

Jander, R. 1970. Ein Einsatz zur modernen Elementarbeschreibung der Orientierungshandlung *Zeitschrift für Tierpsychologie* **27**, 771–8. [P. 71]

Jaynes, J. 1978. In a manner of speaking. *Behavioral and Brain Sciences* **1**, 578–9. [P. 108]

Jennings, H. S. 1906. *Behavior of the lower organisms.* Columbia University Press: New York. [Pp. 93, 99]

Johnstone, J. R. & Mark, R. F. 1969. Evidence for efference copy for eye movements in fish. *Comparative Biochemistry and Physiology* **30**, 931–9. [P. 78]

Jolly, A. 1966. Lemur social behavior and primate intelligence. *Science* **153**, 501–6. [P. 20]

1972. *The evolution of primate behavior.* The Macmillan Co.: New York. [P. 20]

Keeton, W. T. 1967. *Biological science.* W. W. Norton: New York. [Pp. 1, 7, 9, 82, 153]

Kennedy, J. S. 1939. The behaviour of the desert locust (*Schistocerca gregaria* Forsk.) in an outbreak centre. *Transactions of the Royal Entomological Society of London* **89**, 385–542. [P. 7]

1951. The migration of the desert locust (*Schistocerca gregaria* Forsk.). I. The behaviour of swarms. II. A theory of long-range migrations. *Philosophical Transactions of the Royal Society of London, Ser. B* **235**, 163–290. [P. 8]

1954. Is modern ethology objective? *British Journal of Animal Behaviour* **2**, 12–19. [Pp. 34, 35]

1958. The experimental analysis of aphid behaviour and its bearing on current theories of instinct. *Proceedings of the Tenth International Congress of Entomology, Montreal 1956* **2**, 397–404. [Pp. 7, 144, 156]

1966. The balance between antagonistic induction and depression of flight activity in *Aphis fabae* Scopoli. *Journal of Experimental Biology* **45**, 215–28. [Pp. 142, 143]

1967. Behaviour as physiology. Pp. 249–65 in Beament, J. W. L. & Treherne, J. E. (Eds.) *Insects and physiology.* Oliver & Boyd: Edinburgh. [Pp. 130, 144]

1972. The emergence of behaviour. *Journal of the Australian Entomological Society* **11**, 168–76. [P. 8]

1975. Insect dispersal. Pp. 103–19 in Pimental, D. (Ed.) *Insects, science, and society*. Academic Press: New York. [P. 66]

1983. Zigzagging and casting as a programmed response to wind-borne odour: a review. *Physiological Entomology* **8**, 109–20. [P. 148]

1985*a*. Migration, behavioral and ecological. Pp. 5–27 in Rankin (Ed.). [Pp. 64, 65]

1985*b*. Displacement activities and post-inhibitory rebound. *Animal Behaviour* **33**, 1375–7. [Pp. 137, 138]

1986. Some current issues in orientation to odour sources. Pp. 11–25 in Payne, T. L., Birch, M. C. & Kennedy, C. E. J. (Eds.). [Pp. 71, 96, 148, 164]

1987*a*. Animal motivation: the beginning of the end? Pp. 17–31 in Chapman, R. F., Bernays, E. A. & Stoffolano, J. G., Jr. (Eds.) *Perspectives in chemoreception and behavior*. Springer-Verlag: New York. [Pp. 24, 59, 82, 83, 159, 166]

1987*b*. [Review of Toates 1986]. *Animal Behaviour* **35**, 630–2. [Pp. 70, 71]

1990. Behavioural post-inhibitory rebound in aphids taking flight after exposure to wind. *Animal Behaviour* **39**, 1078–88. [P. 142]

Kennedy, J. S. & Booth, C. O. 1963. Coordination of successive activities in an aphid. The effect of flight on the settling responses. *Journal of Experimental Biology* **40**, 351–69. [P. 66]

Kennedy, J. S. & Ludlow, A. R. 1974. Coordination of two kinds of flight activity in an aphid. *Journal of Experimental Biology* **61**, 173–96. [P. 143]

Keynes, G. 1967. Bacon, Harvey and the originators of the Royal Society. *Proceedings of the Royal Society of London, Ser. B* **169**, 1–66. [P. 159]

Klopfer, P. H. 1961. Observational learning in birds: the establishment of behavioral modes. *Behaviour* **17**, 71–80. [Reprinted on pp. 379–84 of Klopfer & Hailman (Eds.).] [P. 48]

Klopfer, P. H. & Hailman, J. P. 1972. (Eds.). *Function and evolution of behavior: an historical sample from the pens of ethologists*. Addison-Wesley Publishing Co.: Reading, MA. [P. 46]

Kovac, M. P. & Davis, W. J. 1977. Behavioral choice: neural mechanisms in *Pleurobranchaea*. *Science* **198**, 632–4. [P. 129]

1980. Reciprocal inhibition between feeding and withdrawal behaviors in *Pleurobranchaea*. *Journal of Comparative Physiology* **139**, 77–86. [P. 129]

Krebs, J. R. 1973. Behavioural aspects of predation. Pp. 73–111 in Bateson, P. P. G. & Klopfer, P. H. (Eds.) *Perspectives in ethology*. Plenum Press: London & New York. [P. 39]

1977. Mental imagery. [Review of Griffin 1976.] *Nature, London* **266**, 792. [Pp. 10, 16, 104]

Krebs, J. R. & Davies, N. B. 1981. *An introduction to behavioural ecology*. Blackwell Scientific Publications: Oxford. [P. 14, 57, 150]

(Eds.) 1984. *Behavioural ecology: an evolutionary approach*. 2nd edn. Blackwell Scientific Publications: Oxford. [P. 61]

1987. *An introduction to behavioural ecology*. 2nd edn. Blackwell Scientific Publications: Oxford. [Pp. 39, 50, 52–3, 61, 63, 89, 150, 152]

Krebs, J. R. & Dawkins, R. 1984. Animal signals; mind-reading and manipulation. Pp. 380–402 in Krebs & Davies (Eds.). [Pp. 56, 158]

Krebs, J. R. & Horn, G. (Eds.) 1990. Behavioural and neural aspects of learning and memory. *Philosophical Transactions of the Royal Society of London, Ser. B* **329**, 97–227. [P. 156]

Kühn, A. 1919. *Die Orientierung der Tiere im Raum*. Gustav Fischer: Jena. [P. 71]

Kummer, H. 1982. Social knowledge in free-ranging primates. Pp. 113–50 in Griffin. [P. 152]

Kummer, H. & Goodall, J. 1985. Conditions for innovative behaviour in primates. *Philosophical Transactions of the Royal Society of London, Ser. B* **308**, 203–14. [P. 48]

Lawrence, E. S. 1985*a*. Evidence for search image in blackbirds (*Turdus morula* L.): short-term learning. *Animal Behaviour* **33**, 929–37. [P. 39]

1985*b*. Evidence for search image in blackbirds (*Turdus morula* L.): long-term learning. *Animal Behaviour* **33**, 1301–9. [P. 39]

1986. Can great tits (*Parus major*) acquire search images? *Oikos* **47**, 3–12. [P. 39]

Lawrence, E. S. & Allen, J. A. 1983. On the term 'search image'. *Oikos* **40**, 313–14. [P. 39]

Lehrman, D. S. 1953. A critique of Konrad Lorenz's theory of instinctive behavior. *Quarterly Review of Biology* **28**, 337–63. [P. 34]

Limber, J. 1980. Language in child and chimp? Pp. 197–220 in Sebeok and Umiker-Sebeok. (Eds.) [Reprinted from *American Psychologist* (1977) **32**, 280–95.] [P. 41]

Lockery, S. 1989. Narrow intentions. Pp. 185–93 in Montefiore & Noble (Eds.). [P. 152]

Lockwood, R. 1985/6. Anthropomorphism is not a four-letter word. Pp. 185–99 in Fox, M. W. & Mickley, L. D. (Eds.) *Advances in animal welfare*. The Humane Society of the United States: Washington, DC. [Pp. 93, 122]

Loeb, J. 1900. *Comparative physiology of the brain and comparative psychology*. G. P. Putnam's Sons: New York & London. [P. 2]

1918. *Forced movements, tropisms, and animal conduct*. Lippincott: New York & London. [Pp. 71, 99]

Lonnendonker, U. & Scharstein, H. 1991. Fixation and optomotor response of walking colorado beetles: interaction with spontaneous turning tendencies. *Physiological Entomology* **16**, 65–76. [P. 79]

Lorenz, K. Z. 1937. Ueber den Begriff der Instinkthandlung. *Folia biotheoretica* **2**, 17–50. [P. 33]

1950. The comparative method in studying innate behaviour patterns. *Symposia of the Society for Experimental Biology* **4**, 221–68. [Pp. 14, 33, 51, 81, 158]

1981. *The foundations of ethology*. Springer-Verlag: Berlin. [P. 33]

Lowe, C. F. 1983. Radical behaviorism and human psychology. Pp. 71–93 in Davey, G. C. L. (Ed.) *Animal models of human behaviour*. John Wiley & Sons: Chichester, Sussex. [Pp. 2, 98]

Ludlow, A. R. 1980. The evolution and simulation of a decision maker. Pp. 274–96 in Toates & Halliday (Eds.). [Pp. 136, 140]

McCleery, R. H. 1983. Interactions between activities. Pp. 134–67 in Halliday & Slater, vol. I. [Pp. 136, 137, 143]

1989. [Review of Colgan 1989.] *Animal Behaviour* **39**, 1091–2. [Pp. 55, 71]

Macdonald, D. W. & Dawkins, M. 1981. Ethology – the science and the tool. Pp. 203–23 in Sperlinger (Ed.). [P. 128]

McFarland, D. J. 1966*a*. On the causal and functional significance of displacement activities. *Zeitschrift für Tierpsychologie* **23**, 217–35. [P. 136]

1966*b*. The role of attention in the disinhibition of displacement activity. *Quarterly Journal of Experimental Psychology* **18**, 19–30. [P. 137]

1969. Mechanisms of behavioural inhibition. *Animal Behaviour* **17**, 238–42. [Pp. 136, 138]

1970. Adjunctive behaviour in feeding and drinking situations. *Revue du Comportement Animal* **4**, 64–73. [P. 143]

1971. *Feedback mechanisms in animal behaviour*. Academic Press: London. [P. 67]

(Ed.) 1981. *The Oxford companion to animal behaviour*. Oxford University Press: Oxford. [Pp. 50, 72, 77]

1985. *Animal Behaviour. Psychobiology, ethology and evolution*. Pitman Publishing Co.: London. [Pp. 39, 61, 92, 98, 136, 137, 138, 142, 149]

1986. Orientation. Pp. 111–13 in Harré & Lamb. [P. 77]

1989*a*. *Problems of animal behaviour*. Longmans: London. [Pp. 2, 22, 27, 29, 30, 31, 54, 72, 73, 74, 98, 133, 163]

1989*b*. Goals, no-goals and own goals. Pp. 39–57 in Montefiore & Noble (Eds.). [Pp. 22, 30, 87, 163]

1989*c*. The teleological imperative. Pp. 211–28 in Montefiore & Noble (Eds.). [Pp. 14, 24, 25, 29(2), 54, 74, 93, 158, 164]

1989*d*. Swan song of a phoenix. Pp. 283–301 in Montefiore & Noble (Eds.). [P. 30]

McFarland, D. J. & Houston, A. I. 1981. *quantitative ethology. The state space approach*. Pitman Advanced Publishing Program: Boston, London & Melbourne. [P. 67]

McGinn, C. 1987. Could a machine be conscious? Pp. 279–88 in Blakemore & Greenfield. (Eds.). [P. 2]

MacKay, D. M. & Mittelstaedt, H. 1974. Visual stability and motor control (reafference revisited). 5. *Kongress der deutschen Gesellschaft für Kybernetik, Nürnberg 1973*, 71–80. [P. 79]

Manning, A. 1979. *An introduction to animal behaviour*. 3rd edn. Edward Arnold: London. [Pp. 54, 142, 167]

1989. Konrad Lorenz and Niko Tinbergen: an appreciation. *Association for the Study of Animal Behaviour Bulletin* **7**, 3–5. [Pp. 34, 58]

Marcel, A. J. & Bisiach, E. (Eds.) 1988. *Consciousness in contemporary science*. Clarendon Press: Oxford.

Marler, P. & Hamilton, W. J. III. 1966. *Mechanisms of animal behavior*. John Wiley & Sons: New York. [Pp. 141, 142]

Marsh, D., Kennedy, J. S. & Ludlow, A. R. 1978. An analysis of zigzagging flight in moths. *Physiological Entomology* **3**, 221–40. [Corrected in *Physiological Entomology* **6**, 225]. [P. 148]

Marshall, J. C. 1982. A la representation du temps perdu. *Behavioral and Brain Sciences* **5**, 382–3. [P. 13]

Medawar, P. B. 1969*a*. *The art of the soluble*. Penguin Books Ltd: Harmondsworth, Mddx. [Pp. 159, 160]

1969*b*. *Induction and intuition in scientific thought*. Methuen & Co.: London. [Pp. 124, 159]

1976. Does ethology throw any light on human behaviour? Pp. 497–506 in Bateson & Hinde (Eds.). [Pp. 12, 18, 22–3]

Mellgren, R. L. (Ed.) 1983. *Animal cognition and behaviour*. North-Holland Publishing Co.: Amsterdam. [Pp. 96, 101]

Menzel, E. W., Savage-Rumbaugh, E. S. & Lawson, J. 1985. Chimpanzee (*Pan troglodytes*) spatial problem solving with the use of mirrors and televised equivalents of mirrors. *Journal of Comparative Psychology* **99**, 211–17. [Pp. 111, 112]

Midgley, M. 1985. Persons and non-persons. Pp. 52–62 in Singer (Ed.). [P. 123]

Mitchell, R. W. 1986. A framework for discussing deception. Pp. 3–40 in Mitchell & Thompson (Eds.). [Pp. 12, 105, 162]

Mitchell, R. W. & Thompson, N. S. (Eds.) 1986. *Deception. Perspectives on human and nonhuman deceit.* State University of New York Press: Albany, NY. [P. 156]

Mittelstaedt, H. 1949. Telotaxis und Optomotorik von *Eristalis tenax* bei Augeninversion. *Naturwissenschaften* **36**, 90. [P. 75]

1958. Regelung in der Biologie. *Regelstechnik* **2**, 177–81. [P. 67]

1964. Basic control patterns in orientational homeostasis. *Symposia of the Society for Experimental Biology* **18**, 364–85. [Pp. 67, 74]

1971. Reafferenzprinzip – Apologie und Kritik. In Keidel, W. D. & Plattig, K. H. (Eds.) *Erlanger Physiologentagung 1970.* Springer-Verlag: Berlin, Heidelberg & New York. [Reprinted in *Ergebnisse der Biologie* **26**, 253–8.]. [P. 79]

1978. Kurs- und Lageregelung. Kybernitische Analyse von Orientierungsleistungen. 6. *Kongress der Deutschen Gesselschaft für Kybernetik 1977*, München. [P. 67]

Miyadi, D. 1964. Social life of Japanese monkeys. *Science* **143**, 783–6. [P. 46]

Montefiore, A. & Noble, D. (Eds.) 1989. *Goals, no-goals and own goals. A debate on goal-directed and intentional behaviour.* Unwin Hyman: London. [P. 30]

Morgan, C. L. 1890. *Animal life and intelligence.* Edward Arnold: London. [P. 19]

1894. *Introduction to comparative psychology.* Scott: London. [P. 153]

Morse, D. H. 1980. Ecological aspects of some mixed species foraging flocks of birds. *Ecological Monographs* **40**, 119–68. [P. 39]

Murton, R. K. 1971. The significance of a specific search image in the feeding behaviour of the wood pigeon. *Behaviour* **40**, 10–41. [P. 39]

Noble, D. 1989. Intentional action and physiology. Pp. 81–100 in Montefiore & Noble (Eds.). [P. 22]

Nottingham, S. F. & Hardie, J. 1989. Migratory and targeted flight in seasonal forms of the black bean aphid, *Aphis fabae*. *Physiological Entomology* **14**, 451–8. [P. 66]

Palameta, B. & Lefebvre, L. 1985. The social transmission of a food-finding technique in pigeons: what is learned? *Animal Behaviour* **33**, 892–6. [P. 49]

Parker, S. T. & Gibson, K. R. (Eds.). 1990. "Language" and intelligence in monkeys and apes. Comparative developmental perspectives. Cambridge University Press: Cambridge.

Passingham, R. E. 1982. *The human primate.* W. H. Freeman & Co.: Oxford & San Francisco. [Pp. 21, 105]

1985. Cortical mechanisms and cues for action. Pp. 101–13 in Weiskrantz, L. (Ed.). Animal intelligence. *Philosophical Transactions of the Royal Society of London, Ser. B* **308**, 1–216. [P. 126]

1989. The origins of human intelligence. Pp. 123–36 in Durant. [Pp. 16, 18]

Paul, D. H. 1989. Non-spiking stretch receptors of the crayfish swimmeret receive an efference copy of the central motor pattern for the swimmeret. *Journal of Experimental Biology* **141**, 257–64. [P. 78]

Payne, T. L., Birch, M. C. & Kennedy, C. E. J. (Eds.) 1986. *Mechanisms in insect olfaction.* Clarendon Press: Oxford.

Penrose, R. 1987. Minds, machines and mathematics. Pp. 259–76 in Blakemore & Greenfield (Eds.). [Pp. 2, 20]

Pietrewicz, A. T. & Kamil, A. C. 1979. Search image formation in the blue jay (*Cyanocitta cristata*). *Science* **204**, 1332–3. [P. 39]

1981. Search images and the detection of cryptic prey: an operant approach. Pp. 311–31 in Kamil, A. C. & Sargent, T. D. (Eds.) *Foraging behaviour. Ecological, ethological and psychological approaches.* Garland STPM Press: New York. [P. 39]

Rankin, M. A. (Ed.) 1985. Migration: mechanisms and adaptive significance. *Contributions in Marine Science* **27** Supplement. [P. 63]

Restle, F. 1957. Discrimination of cues in mazes. A resolution of the "Place vs. Response" question. *Psychological Review* **64**, 217–28. [P. 102]

Ridley, M. 1982. Tasteless behaviour. [Review of Lorenz 1981]. *Nature, London* **295**, 439–40. [P. 33]

1986. *Animal behaviour: a concise introduction.* Blackwell Scientific Publications: Oxford. [Pp. 39, 50, 55, 61, 142, 150]

Ristau, C. A. & Robbins, D. 1982. Language in the great apes: a critical review. *Advances in the Study of Behaviour* **12**, 141–255. [P. 44]

Robinson, D. L. & Wurtz, R. H. 1976. Use of an extraretinal signal by monkey superior colliculus neurons to distinguish real from self-induced stimulus movement. *Journal of Neurophysiology* **39**, 852–70. [P. 78]

Roeder, K. D. 1965. Epilogue. Pp. 247–52 in Treherne, J. E. & Beament, J. W. L. (Eds.) *Physiology of the insect nervous system.* Academic Press: London. [P. 57]

Roitblat, H. L. 1987. *Introduction to comparative cognition.* W. H. Freeman: New York. [Pp. 96, 104]

Roitblat, H. L., Bever, T. G. & Terrace, H. S. (Eds.) 1984. *Animal cognition.* Lawrence Erlbaum: Hillsdale, NJ. [P. 96]

Rollin, B. E. 1989. *The unheeded cry. Animal consciousness, animal pain and science.* Oxford University Press: Oxford. [Pp. 116, 120]

Romanes, G. J. 1882. *Animal intelligence.* Kegan Paul, Trench: London. [P. 15]

1883. *Mental evolution in animals.* Kegan Paul, Trench: London. [P. 15]

Rooijen, J. van. 1981. Are feelings adaptations? The basis of modern animal ethology. *Applied Animal Ethology* **7**, 187–9. [P. 114]

Roper, T. J. 1983. Schedule-induced behaviour. Pp. 127–55 in Mellgren (Ed.). [P. 143]

1984. Response of thirsty rats to absence of water: frustration, disinhibition or compensation? *Animal Behaviour* **32**, 1225–35. [Pp. 138, 139]

1985. How plausible is post-inhibitory rebound as an account of displacement activity? A reply to Kennedy. *Animal Behaviour* **33**, 1377–8. [P. 142]

Roper, T. J. & Crossland, G. 1982. Mechanisms underlying eating-drinking transitions in rats. *Animal Behaviour* **30**, 602–14. [Pp. 136, 138]

Roughgarden, J., May, R. M. & Levin, S. A. 1989. *Perspectives in ecological theory.* Princeton University Press: Princeton, NJ. [P. 61]

Royama, T. 1970. Factors governing the hunting behaviour and selection of food by the great tit, *Parus major. Journal of Animal Ecology* **39**, 619–68. [P. 38]

Ruiter, L. de 1952. Some experiments on the camouflage of stick caterpillars. *Behaviour* **4**, 222–32. [P. 37]

Rumbaugh, D. M. (Ed.) 1977. *Language learning by a chimpanzee: the Lana project.* Academic Press: New York. [P. 45]

Russell, E. S. 1934. *The behaviour of animals*. Edward Arnold: London. [P. 4]
1946. *The directiveness of organic activities*. Cambridge University Press: Cambridge. [P. 4]
Savage-Rumbaugh, E. S. 1986. *Ape language: from conditioned response to symbol*. Colombia University Press: New York. [Pp. 43, 44, 110, 111, 113]
Savage-Rumbaugh, E. S., Rumbaugh, D. M. & Boysen, S. 1978. Symbolic communication between two chimpanzees (*Pan troglodytes*). *Science* **201**, 641–4. [P. 66]
1980. Do apes use language? *Americal Scientist* **68**, 49–61. [P. 43]
Schleidt, W. M. 1981. The behaviour of organisms, as it is linked to genes and populations. *Perspectives in Ethology* **4**, 147–56. [P. 125]
Schmidt-Koenig, K. 1971. *Avian orientation and navigation*. Academic Press: New York. [P. 65]
Schöne, H. 1984. *Spatial orientation. The spatial control of behaviour in animals and man*. Princeton University Press: Princeton, NJ. [Pp. 71, 74, 80, 147, 153]
Sebeok, T. A. & Umiker-Sebeok, J. (Eds.). 1980. *Speaking of apes. A critical anthology of two-way communication with man*. Plenum Press: New York. [P. 40, 41, 45]
Seyfarth, R. M. 1982. Communication as evidence of thinking. State of the art report. Pp. 391–406 in Griffin (Ed.). [P. 43]
Selverston, A. (Ed.) 1985. *Model neural networks and behavior*. Plenum Press: New York & London. [P. 142]
Sevenster, P. 1961. A causal analysis of a displacement activity (fanning in *Gasterosteus aculeatus*). *Behaviour Supplement* **9**, 1–170. [P. 136]
Sherrington, C. S. 1913. Reflex inhibition as a factor in the coordination of movements and postures. *Quarterly Journal of Experimental Physiology* **6**, 251–310. [P. 139]
1947 [reprinted from 1906]. *The integrative action of the nervous system*. Cambridge University Press: Cambridge. [Pp. 86, 87, 142]
Sherry, D. F. & Galef, B. G., Jr 1984. Cultural transmission without imitation: milk bottle opening by birds. *Animal Behaviour* **32**, 937–8. [P. 48]
1990. Social learning without imitation: more about milk bottle opening by birds. *Animal Behaviour* **40**, 987–9. [P. 48]
Shettleworth, S. J. & Mrosovsky, N. 1990. From one subjectivity to another. *Behavioral and Brain Sciences* **13**, 37–8. [Pp. 119, 122]
Shorey, H. H. 1973. Behavioural responses to insect pheromones. *Annual review of Entomology* **18**, 349–80. [Pp. 145, 147]
1976. *Animal communication by pheromones*. Academic Press: New York & London. [Pp. 145, 147]
Singer, P. (Ed.) 1985. *In defence of animals*. Basil Blackwell: Oxford. [P. 120]
Skinner, B. F. 1938. *The behavior of animals: an experimental analysis*. Appleton-Century-Crofts: New York. [P. 2]
Slater, P. J. B. 1985. *An introduction to ethology*. Cambridge University Press: Cambridge. [Pp. 61, 167]
Slobodkin, L. B. 1977. Evolution is no help. *World Archaeology* **8**, 332–43. [P. 105]
Smith, J. N. M. & Dawkins, R. 1971. The hunting behaviour of individual great tits in relation to spatial variations in their food density. *Animal Behaviour* **19**, 695–706. [P. 38]
Smith, W. J. 1986. An "informational" perspective on manipulation. Pp. 71–86 in Mitchell & Thompson (Eds.). [Pp. 22, 158]

Sperlinger, D. (Ed.) 1981. *Animals in research: new perspectives in animal experimentation.* John Wiley: Chichester & New York. [P. 115]

Sperry, R. W. 1950. Neural basis of the spontaneous optokinetic response produced by visual inversion. *Journal of Comparative and Physiological Psychology* **43**, 482–9. [P. 77]

Staddon, J. E. R. 1977. Scheudule-induced behavior. Pp. 125–52 in Honig, W. K. & Staddon, J. E. R. (Eds.) *Handbook of operant behavior.* Prentice-Hall: Englewood Cliffs, NJ. [P. 143]

1983. *Adaptive behavior and learning.* Cambridge University Press: Cambridge. [Pp. 59, 128, 142, 166]

1986. The comparative psychology of operant behavior. Pp. 83–94 in Lowe, C. E., Richelle, M. & Blackman, D. E. (Eds.) *Behavior analysis and contemporary psychology.* Lawrence Erlbaum: Hillsdale, NJ. [P.153]

1989. Animal psychology: the tyranny of anthropocentrism. *Perspectives in Ethology* **8**, 123–36. [Pp. 6, 102]

Stillar, K. T. 1985. Comparative overview and perspectives. *Society for Experimental Biology Symposium Series* **24**, 303–16. [P. 76]

Stuart, R. J. 1983. A note on terminology in animal behaviour with special reference to slavery in ants. *Animal Behaviour* **31**, 1259–60. [Pp. 24, 154]

Swingland, I. 1984. [Review of Baker 1982.] *Animal Behaviour* **32**, 310. [P. 61]

Szentágothai, J. 1987. The 'brain–mind' relation: a pseudoproblem? Pp. 323–6 in Blakemore & Greenfield (Eds.). [P. 124]

Tavolga, W. N. 1969. *Principles of Animal Behavior.* Harper & Row: New York. [P. 124]

Terrace, H. S. 1979. *Nim: a chimpanzee who learned sign language.* A. A. Knopf: New York. [Pp. 1, 42]

1984*a*. 'Language' in apes. Pp. 179–207 in Harré & Reynolds (Eds.). [Pp. 42, 43]

1984*b*. Animal cognition. Pp. 7–28 in Roitblat *et al.* (Eds.). [P. 101]

1984*c*. A behavioral theory of mind? *Behavioral and Brain Sciences* **7**, 469–71. [Pp. 98, 101]

1985. Animal cognition: thinking without language. *Philosophical Transactions of the Royal Society of London, Ser. B* **200**, 113–28. [P. 96]

1986. Foreword. Pp. xiii-xx in Savage-Rumbaugh. [Pp. 43, 44]

Terrace, H. S., Petitto, L. A., Sanders, R. J. & Bever, T. G. 1979. Can an ape create a sentence? *Science* **200**, 891–902. [P. 41]

Teuber, H. L. (Chairman) 1974. Panel discussion. Key problems in the programming of movements. *Brain Research* **71**, 533–68. [P. 76]

Thompson, C. R. & Church, R. M. 1980. An explanation of the language of a chimpanzee. *Science* **208**, 313–14. [P. 45]

Thorpe, W. H. 1948. The modern concept of instinctive behaviour. *British Journal of Animal Behaviour* **1**, 1–12. [P. 33]

1951. The learning abilities of birds. Part 2. *Ibis* **93**, 252–96. [P. 46]

1954. Some concepts of ethology. *Nature* **174**, 101–5. [P. 33]

1956. Some implications of the study of animal behaviour. *Advancement of Science* **13**, 42–55. [P. 36]

1963. *Learning and instinct in animals.* 2nd Edn. Methuen & Co.: London. [Pp. 4, 36, 46, 48]

1965. Ethology and consciousness. Pp. 470–505 in Eccles, J. C. (Ed.) *Brain and conscious experience.* Springer-Verlag: New York. [P. 4]

Tinbergen, L. 1960. The natural control of insects in pine woods. I. Factors influencing the intensity of predation by songbirds. *Archives Néerlandaises de Zoologie* **13**, 265–336. [P. 37]

Tinbergen, N. 1942. An objectivistic study of the innate behaviour of animals. *Bibliotheca biotheoretica* Leiden **1**, 40–98. [Pp. 34, 52, 151]

1950. The hierarchical organisation of nervous mechanisms underlying instinctive behaviour. *Symposia of the Society for Experimental Biology* **4**, 305–12. [Pp. 33, 55, 124, 141]

1951. *The study of instinct*. Clarendon Press: Oxford. [Pp. 9, 16, 24, 28, 33, 35, 38, 49, 51, 54, 59, 90, 151, 166]

1952. "Derived" activities. Their causation, biological significance, origin and emancipation during evolution. *Quarterly Review of Biology* **27**, 1–32. [Pp. 135, 138]

1954. Some neurophysiological problems raised by ethology. *British Journal of Animal Behaviour* **2**, 115. [Pp. 59, 126, 128, 133]

1963. On aims and methods in ethology. *Zeitschrift für Tierpsychologie* **20**, 410–33. [Pp. 34, 55, 135, 151, 165, 166]

1965. Behavior and natural selection. *Proceedings of 16th International Congress of Zoology, Washington* **6**, 521–42. [P. 156]

1969. Ethology. Pp. 238–68 in Harre, R. (Ed.) *Scientific thought 1900–1960*. Clarendon Press: Oxford. [Pp. 34, 59, 76, 124, 125]

Toates, F. M. 1980. *Animal behaviour. A systems approach*. John Wiley & Sons: Chichester, Sussex. [Pp. 67, 142]

1984*a*. The behaviourist approach to motivation and learning – a personal view. *New Psychologist* February, 40–9. [Pp. 10, 68, 69]

1984*b*. Models, yes; homunculus, no. *Behavioral and Brain Sciences* **7**, 650–1. [P. 68]

1986. *Motivational systems*. Cambridge University Press: Cambridge. [Pp. 67, 68, 69, 102, 103]

1990. Broadening the welfare index. *Behavioral and Brain Sciences* **13**, 40–1. [P. 120]

Toates, F. M. & Birke, L. 1982. Motivation: a new perspective on some old ideas. *Perspective in Ethology* **5**, 191–241. [P. 56, 57]

Toates, F. M. & Halliday, T. R. (Eds.) 1980. *Analysis of motivational processes*. Academic Press: London. [P. 67]

Tobin, T. R. 1981. Pheromone attraction; role of internal control mechanisms. *Science* **214**, 1147–9. [P. 148]

Tolman, E. C. 1948. Cognitive maps in rats and men. *Psychological Review* **55**, 189–208. [P. 101]

Traynier, R. M. M. 1968. Sex attraction in the Mediterranean flour moth, *Anagasta kuhniella*: location of the female by the male. *Canadian Entomologist* **100**, 5–10. [P. 103]

Trivers, R. L. 1971. The evolution of reciprocal altruism. *Quarterly Review of Biology* **46**, 35–57. [P. 21]

1985. *Social evolution*. Benjamin/Cummings Publishing Co.: Menlo Part, CA. [Pp. 21, 22]

Tudge, C. 1987 More than simple machines [Review of Evans.] *New Scientist* 23 July, 51. [P. 154]

Uexkull, J. von. 1934. *Streifzüge durch die Umwelten von Tieren und Menschen*. Springer: Berlin. [P. 37]

Umiker-Sebeok, J. & Sebeok, T. A. 1981. Clever Hans and smart simians: the self-fulfilling prophecy and methodological pitfalls. *Anthropos* **76**, 89–165. [P. 40]

Van Gulick, R. 1988. Consciousness, intrinsic intentionality, and self-understanding machines. Pp. 78–100 in Marcel & Bisiach (Eds.). [P. 2]

Visalberghi, E. & Fragaszy, D. M. 1990*a*. Food washing behaviour in tufted capuchin monkeys, *Cebus apella*, and crabeating macaques, *Macacus fascicularis*. *Animal Behaviour* **40**, 829–36. [Pp. 26, 49]

1990*b*. Do monkeys ape? Pp. 247–73 in Parker & Gibson (Eds.). [P. 48]

Waal, F. de 1986. Deception in the natural communication of chimpanzees. Pp. 221–44 in Mitchell & Thompson (Eds.). [P. 27]

1989. *Peacemaking among primates*. Harvard University Press: Cambridge, MA. [P. 88]

Wall, P. D. 1985. Future trends in pain research. *Philosophical Transactions of the Royal Society of London, Ser. B* **308**, 393–401. [Pp. 117, 118]

Wallace, R. A. 1983. *The ecology and evolution of animal behaviour*. Goodyear Publishing Co.: Pacific Palisades, CA. [P. 39]

Washburn, M. F. 1926. *The animal mind*. 3rd Edn. Macmillan: New York. [Pp. 4, 24, 99]

Wasserman, E. A. 1984. Animal intelligence: understanding the minds of animals through their behavioural "ambassadors". Pp. 45–60 in Roitblat, Bever & Terrace (Eds.). [Pp. 15, 99]

Watson, J. B. 1930. *Behaviorism*. Revised Edn. W. W. Norton: New York. [P. 2]

Wehner, R. & Menzel, R. 1990. Do insects have cognitive maps? *Annual Review of Neuroscience* **13**, 403–14. [P. 101]

Weiskrantz, L. 1987. Neuropsychology and the nature of consciousness. Pp. 307–20 in Blakemore & Greenfield (Eds.). [Pp. 13, 19, 20, 30, 125]

Whiten, A. 1989. Transmission mechanisms in primate cultural evolution. *Trends in Ecology and Evolution* **4**, 61–2. [Pp. 22, 46, 48]

Wilder, H. 1990. Interpretive cognitive ethology, Pp. 344–68 in Bekoff & Jamieson (Eds.) 1990*a*. [Pp. 89, 98, 100, 101]

Wilkes, K. V. 1988. –, yishi, duh, um, and consciousness. Pp. 16–41 in Marcel & Bisiach (Eds.). [P. 161]

Wilson, E. O. 1975. *Sociobiology. The new synthesis*. The Belknap Press, Harvard University Press: Cambridge, MA. [P. 48]

Wirtshafter, D. & Davis, J. D. 1977. Set points, settling points, and the control of body weight. *Physiology and Behavior* **19**, 75–8. [P. 74]

Wood-Gush, G. D. M., Dawkins, M. & Ewbank, R. (Eds.) 1981. *Self-awareness in domesticated animals*. Universities Federation for Animal Welfare: South Mimms, Herts. [P. 115]

Wright, R. H. 1958. The olfactory guidance of flying insects. *Canadian Entomologist* **90**, 81–9. [P. 103]

1962. How insects follow a scent. *New Scientist* **14**, 339–41. [P. 103]

1964. *The science of smell*. Allen & Unwin: London. [P. 103]

Young, D. 1989. *Nerve cells and animal behaviour*. Cambridge University Press: Cambridge. [P. 58]

Zalucki, M. P. & Kitching, R. L. 1982. The analysis and description of movement in adult *Danaus plexippus* (Lepidoptera: Danainae). *Behaviour* **80**, 174–98. [P. 66]

Zaretsky, M. & Fraser-Rowell, C. H. 1979. Saccadic suppression by corollary discharge. *Nature, London* **280**, 583–4. [P. 78]

Zihlman, A. L. 1989. Common ancestors and uncommon apes. Pp. 81–105 in Durant (Ed.). [P. 20]

SUBJECT INDEX

SUBJECT INDEX

SUBJECT INDEX

SUBJECT INDEX